Wavelet Radio

Adaptive and Reconfigurable Wireless Systems Based on Wavelets

The first book to provide a detailed discussion of the application of wavelets in wireless communications, this is an invaluable source of information for graduate students, researchers, and telecommunications engineers, managers and strategists. It surveys applications, explains how to design new wavelets, and compares wavelet technology with existing OFDM technology.

- Addresses the applications and challenges of wavelet technology for a range of wireless communication domains
- Aids in the understanding of wavelet packet modulation and compares it with OFDM
- Includes tutorials on convex optimisation and spectral factorisation for the design of wavelets
- Explains design methods for new wavelet technologies for wireless communications, addressing many challenges, such as peak-to-average power ratio reduction, interference mitigation, reduction of sensitivity to time, frequency and phase offsets, and efficient usage of wireless resources
- Describes the application of wavelet radio in spectrum sensing of cognitive radio systems.

Homayoun Nikookar is an Associate Professor in the Faculty of Electrical Engineering, Mathematics, and Computer Science at Delft University of Technology, where he leads the Radio Advanced Technologies and Systems (RATS) programme, and supervises a team of researchers carrying out cutting-edge research in the field of advanced radio transmission. He has received several paper awards at international conferences and symposiums and the "Supervisor of the Year Award" at Delft University in 2010.

EuMA High Frequency Technologies Series

Series Editor
Peter Russer, Technical University of Munich

Forthcoming

Thomas Zwick et al. (Eds), *Ultra Wideband RF System Engineering*
Peter Russer, Andreas Cangellaris and Uwe Siart, *Interference and Noise in Electromagnetics*
Maurizio Bozzi, Apostolos Georgiadis and Ke Wu, *Substrate Integrated Waveguides*
Er-Ping Li and Hong-Son Chu, *Plasmonic Nanoelectronics and RF Nanotechnology*
Luca Roselli (ed), *Green RFID Systems*
George Deligeorgis, *Graphene Device Engineering, High Frequency and Sensing Applications*

Wavelet Radio

Adaptive and Reconfigurable Wireless Systems Based on Wavelets

HOMAYOUN NIKOOKAR

CAMBRIDGE UNIVERSITY PRESS
Cambridge, New York, Melbourne, Madrid, Cape Town,
Singapore, São Paulo, Delhi, Mexico City

Cambridge University Press
The Edinburgh Building, Cambridge CB2 8RU, UK

Published in the United States of America by Cambridge University Press, New York

www.cambridge.org
Information on this title: www.cambridge.org/9781107017801

First published 2013

Printed and bound in the United Kingdom by the MPG Books Group

A catalogue record for this publication is available from the British Library

Library of Congress Cataloguing in Publication data
Nikookar, Homayoun.
Wavelet radio : adaptive and reconfigurable wireless systems based on wavelets / Homayoun Nikookar.
pages cm. – (EuMA high frequency technologies series)
Includes bibliographical references and index.
ISBN 978-1-107-01780-1
1. Wireless communication systems. 2. Radio. 3. Adaptive signal processing. 4. Wavelets (Mathematics) I. Title.
TK5103.2.N55 2013
621.384 – dc23 2012035060

ISBN 978-1-107-01780-1 Hardback

Contents

Preface

Wavelets provide promising potential applications in wireless communication. The main property of wavelets for these applications is their ability and flexibility to characterize signals with adaptive time–frequency resolution. The convergence of information, multimedia, entertainment and wireless communications has raised hopes of realizing the vision of ubiquitous communication. To actualize this, there is a challenge of developing technologies and architectures capable of handling large volumes of data under severe constraints of resources such as power and bandwidth. Wavelets are uniquely qualified to address this challenge. They have strong advantage of being generic schemes whose actual characteristics can be widely customized to fulfil the various requirements and constraints of advanced mobile communications systems. The wavelet technology is the choice for smart and resource aware wireless systems.

In the light of this, the objective of this book is to utilize the wavelet technology for smart and resource aware radio systems and to develop wavelet based radio systems that cleverly and efficiently use available resources to guarantee the required quality of service. Adaptation, smartness, context aware, robustness and reconfigurability are the major accents of wavelet radio, which will be concentrated on in this book. This is actualized by developing a wavelet-packet-based multi-carrier modulation radio that can be adaptively reconfigured to operate under different use cases even while maximizing resource utilization.

In a recent paper in the IEEE Communication magazine, Steve Weinstein [1], a pioneer in the development of OFDM, traces back the journey of OFDM right from its inception in 1966 when Chang [2] published the first paper on multi-carrier modulation, to the development of the first proof of concept by Bell Labs in 1985 [3] and its first major consumer deployment as ADSL in 1993 and finally its standardization as IEEE 802.11a in 1999. And in his concluding remarks he advocates wavelet based systems as true successors of OFDM, especially for the development of futuristic low power "Green Radios" which are intelligent and adaptable.

The research and investigation on the utilization of wavelet technology for smart resource aware radio systems as presented in this book, not only aims at tackling the various technical questions that will shorten the development time from conception to practical realization of wavelet radios (vis-à-vis the OFDM cycle which took close to 35 years), but also to give the wireless radio research community a lead in this exciting new line of research. Furthermore, in an era when bold predictions that the *PHY Layer is*

Dead [4] are made, the work on wavelet radio will increase the capacities of the wireless link and open new vistas for an exciting line of research topics on radio design.

The book addresses the physical layer challenges of wavelet radio transmission and is organized in eight chapters. The material is categorized into three broad divisions, namely, theoretical background (Chapters 1, 2), wavelet radio (Chapters 3–5) and wavelet applications in cognitive radio design (Chapters 6 and 7). Finally, the book rounds off in Chapter 8 with conclusions and recommendations for future research. I would greatly appreciate the readers' comments on this work; collaboration and cooperation will leverage the knowledge of the research community.

References

[1] S.B. Weinstein, "Introduction to the History of OFDM," IEEE Communications Magazine, November 2009, Volume 47, No. 11.

[2] R.W. Chang, "High-Speed Multichannel Data Transmission with Bandlimited Orthogonal Signals," Bell Sys. Tech. J., vol. 45, Dec. 1966, pp. 1775–96; see also U.S. Patent 3,488,445, Jan. 6, 1970.

[3] "A Brief History of OFDM", 2010, Web link: http://www.wimax.com/commentary/wimax_weekly/sidebar-1-1-a-brief-history-of-ofdm.

[4] Panel Discussions at IEEE 69th Vehicular Technology Conference (VTC-Spring), April 2009, Barcelona, Web link: http://www.ieeevtc.org/vtc2009spring/panels.php#Panel02.

Acknowledgement

At the outset I would like to express my deepest gratitude to my former PhD student Dr. M.K. Lakshmanan for his kind co-operation in carrying out a world-class cutting-edge research on wavelet radio. His constant support in providing me with input material for this book including figures and diagrams as well as long and productive discussions with him on wavelet radio research is highly appreciated.

Special thanks go to Prof. L.P. Ligthart (emeritus Professor of TU Delft) for his constant support and for his encouragement and full endorsement of my Radio Advanced Technologies and Systems (RATS) research and education program. A broad program which wavelet radio research is a part of that.

I would like to express my deepest gratitude to Prof. R. Prasad (Director of CTIF, Aalborg University) for his vision and perspectives in future wireless communications giving me the opportunity to deliver parts of this book in several international tutorials, among them at the VITAE2009 conference in Aalborg.

My special thanks are due to Prof. N. Baken from Dutch Telecom company KPN for his cooperation, true vision, inspiration and support. I also thank the Dutch Research Delta (DRD) for funding this research work. I would also like to extend my thankfulness to the many fine members of DRD and especially Professor E. Fledderus (from TNO and TU Eindhoven) for his support, stimulating suggestions and encouragement all through the wavelet radio research project.

I would like to thank Prof. P. Russer from TU Munich for his encouragement in publication of this book in the European Microwave Association (EuMA) Microwave and Wireless book series of the Cambridge University Press. My special thanks go to Dr J. Lancashire from Cambridge University Press for her editorial work in the past years. The author acknowledges with great pleasure the publishing assistance of Ms. M. Balashova from Cambridge University Press.

I would also like to extend my thankfulness to my current PhD students X. Lian and H. Lu, my exchange PhD student V. Roy and and my previous PhD student I. Budiarjo, who contributed to this book in one way or another. I acknowledge with great pleasure interesting discussions with the PhD student X. Tau. The contribution of my former MSc students B. Negash, J. Karamehmedovic, B. Torun, N.M. Tessema, A. Bajpai, R. Mulyawan, P. HariMukti as well as D. Ariananda is highly appreciated.

Finally, I dedicate this work to my loving family (Mahsa, Bardia and Parmida) for without their unwavering and unconditional love, affection, support and sacrifice, I would not have been able to write this book.

Homayoun Nikookar

1 Introduction

1.1 Background

The advancements in the field of digital wireless communication have led to many exciting applications like mobile internet access, healthcare and medical monitoring services, smart homes, combat radios, disaster management, automated highways and factories. With each passing day novel and advanced services are being launched, while existing ones continue to flourish. While traditionally only voice and data communication were possible, wireless services have now found applicability in other sectors too including healthcare, transportation, security, logistics, education and finance. For example, telemedicine can render emergent and easy-to-access healthcare at distance. Through rural connectivity, people living in remote places in developing/underdeveloped nations can be given access to good-quality education via long-distance learning programs. In the era of open course ware (OCW), this can prove to be a boundary breaker in spreading top quality educational content to students who hitherto might not have access to them. Demand for wireless services is thus expected to grow in the foreseeable future.

However, with increasing popularity of the wireless services the requirements on prime resources like battery power and radio spectrum are put under severe pressure. For example, currently most spectrum have been allocated, and it is becoming increasingly difficult to find frequency bands that can be made available either for new services or to expand existing ones. Even as the available frequency spectrum appears to be fully occupied, a survey [1] conducted by the American regulatory body Federal Communications Commission (FCC) in 2002 revealed that much of the available spectrum is underused most of the time. The study [1] showed that only 20% or less of the spectrum is used and that spectrum congestions are more due to the sub-optimal use of spectrum than to the lack of free spectrum.

Studies have also shown that the volume of data is increasing by a factor of 18 in five years [2]. For example, the global mobile data was 0.6 exabytes per month in 2011 and more than doubled in 2012 (1.3 exabytes/month), [2]. This number is expected to grow to 4.2 exabytes in 2014 and about 10.8 exabytes/month in 2016 (see Figure 1.1). The 18-fold increase in data volume in five years corresponds to an increase of the associated energy consumption by more than 20% annually. In fact, the current world-wide energy requirements of information and communication technology (ICT) systems contribute nearly 2% of the CO_2 emissions, a figure comparable with the total emissions due to global air travel or about one quarter of the emissions due to cars and trucks.

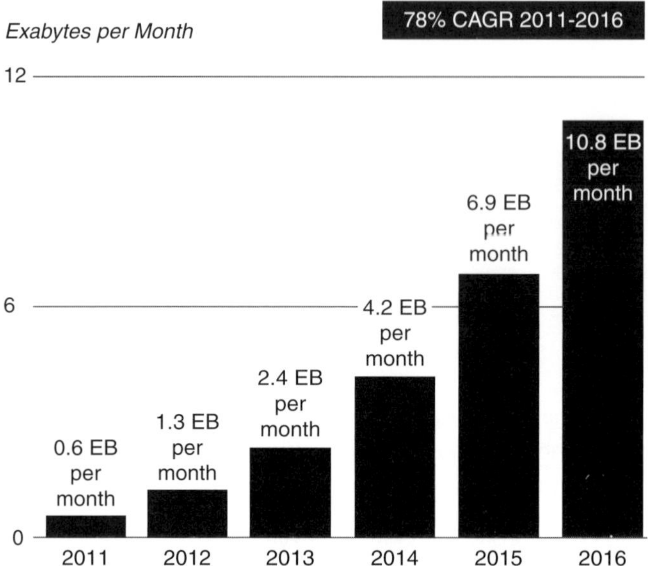

Figure 1.1 Global mobile data traffic growth (mobile traffic will grow by 18 times from 2011 to 2016). Notation in the figure: EB: exabytes (10^{18} bytes), after [2].

Another emerging trend is the demand for higher data rates as exemplified in Figure 1.2, where the growth of home bandwidth since the 1970s has been shown [3]. Today, the Universal Mobile Telecommunications Systems (UMTS) is one of the fastest solutions on the market that can operate in dispersive environments at a rate of 3.84×10^6 chips, but the rapid progress of telecommunication market has created a need for newer techniques that can accommodate data rates even higher than this.

1.1.1 The need

Thus, in a wireless environment the system requirements, network capacities and device capabilities have enormous variations giving rise to significant design challenges. There is therefore an emergent need for developing energy efficient, green technologies that optimize premium radio resources, such as power and spectrum, even while guaranteeing a desirable quality of service. Of signal interest is the development of a capable radio/PHY layer platform that facilitates optimum utilization of energy in addition to guaranteeing spectral efficiency, adequate coverage and good quality of service (QoS). Spatially, temporally and spectrally localized transmission strategies that minimize the energy spent to transmit information-bearing symbols will be crucial towards achieving high energy efficiency. Moreover, wireless systems operate under dynamic conditions with frequent changes in the propagation environment and user requirements. All these trends point to an untapped niche available for flexible, reconfigurable systems that can adapt to its radio neighbourhood.

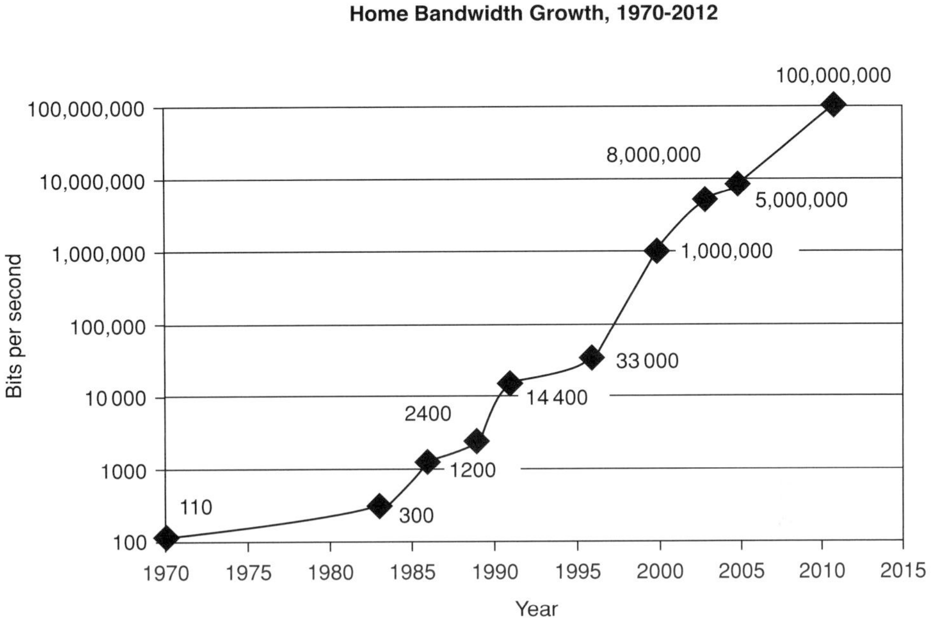

Figure 1.2 Growth of home bandwidth since the 1970s, after [3].

1.1.2 The means

Existing wireless systems and services are based on the mathematical precept of Fourier transform. In comparison to the Fourier transform, the recently formulated theory of wavelets offers many advantages for the design of wireless communications. The main property of wavelets for these applications is their ability to characterize signals with adaptive time–frequency resolution. The goal of this book is to build a generic parameterized baseband radio platform based on wavelet technology as a suitable PHY layer candidate for the design and development of communication systems that can handle large volumes of data under severe constraints of interference and resources such as power and bandwidth. By careful adaptation of the main system parameters according to the radio environment, the operation of the wavelet-based radio is optimized to save valuable energy resources. The system parameterization is realized through a generalized wavelet packet modulator (WPM) that is based on the theory of wavelets and filter banks.

1.2 Wavelet transform as a tool for wireless communications

1.2.1 Wavelets and wavelet transform

A wavelet is a waveform of limited duration. As the name suggests, wavelets are small waveforms with a set of oscillatory structures that is non-zero for a limited period of time (or space). The wavelet transform is a multi-resolution analysis mechanism where an input signal is decomposed into different frequency components with each component studied with resolutions matched to its time-scales. The Fourier transform also decomposes signals into elementary waveforms, but these basis functions are sines and cosines. Thus, when one wants to analyze the local properties of the input signal, such as edges or transients, the Fourier transform is not an efficient analysis tool. By contrast, the wavelet transforms that use irregularly shaped wavelets offer better tools to represent sharp changes and local features.

The wavelet transform is used in various applications and is finding tremendous popularity among technologists, engineers and mathematicians alike. In most of the applications, the power of the transform comes from the fact that the basis functions of the transform are localized in time (or space) and frequency, and offer different resolutions in these domains. These resolutions often correspond to the natural behaviour of the process one wants to analyze, hence the power of the transform. Such properties make wavelets and wavelet transforms natural choices in fields as diverse as image synthesis, data compression, computer graphics and animation, human vision, radar, optics, astronomy, acoustics, seismology, nuclear engineering, biomedical engineering, magnetic resonance imaging, music, fractals, turbulence and pure math.

While the wavelet transform is the *de jure* standard[1] for many signal-processing applications including the fields of image processing, speech processing and data compression, the technique has very rarely been applied to the design of communication systems. This lacuna in existing knowledge is in part a motivation for this book.

1.2.2 Advantages of wavelet transform for wireless communication

The motivation for pursuing wavelet based systems primarily lies in the freedom they provide to communication systems designers [4], [5]. Unlike the Fourier bases that are static sines/cosines, wavelet bases offer flexibility and adaptation that can be tailored to satisfy an engineering demand. This feature is attributable to the fact that the wavelet transform is implemented entirely using filter bank tree structures obtainable from paraunitary filters[2]. The freedom to alter the properties of the wavelet and the filter bank tree structure gives the opportunity to fine-tune and optimize the modulated signal according to application at hand.

The benefits of wavelet-based radios for research and development of energy-efficient communication are summarized as follows:

1.2.2.1 Intelligent utilization of signal space

The wavelet-based systems are realized from tree structures obtained by cascading a fundamental quadrature mirror filter (QMF) pair of low- and high-pass filters. The construction of this tree structure can be adjusted to produce an optimum tree structure that caters to various requirements. The requirements could typically be:

[1] Examples include JPEG 2000, an image compression standard, and MPEG-4 Part 14 or MP4, a multimedia container format standard.

[2] Paraunitary filters are a class of perfect-reconstruction filters that generate orthogonal bases.

- identification and isolation of time–frequency "atoms" affected by an interfering source and communicating around the source of interference [6];
- flexibility with time–frequency tiling of the carriers that can lead to multi-rate systems which can transmit with different rates in different bands [7], a feature that can be exploited in scenarios where the channel characteristics are not uniform.

1.2.2.2 Design of wavelets to customize transceiver characteristics

By careful selection of the fundamental filters, which greatly influence the transmission characteristics, it is possible to optimize the system performance in terms of the bandwidth efficiency, localization of the transmitted signal in time and frequency, minimization of inter-symbol interference (ISI), inter-carrier interference (ICI) or peak-to-average-power ratio (PAPR), robustness towards interference from competing sources. This can also aid in opportunistic communication (e.g., cognitive radio) where unused resources can be cleverly utilized.

1.2.2.3 Flexibility with sub-carriers

The derivation of wavelets is directly related to the iterative nature of the wavelet transform. The wavelet transform allows for a configurable transform size and hence a configurable number of carriers. This facility can be used, for instance, to reconfigure a transceiver according to a given communication protocol; the transform size could be selected according to the channel impulse response characteristics, computational complexity or link quality [6].

1.2.2.4 Enhanced multi-access transmission

Wavelets offer a new dimension of diversity called the "waveform diversity" that can be exploited to enhance multiple access transmission [8]. The wavelet transform generates wavelet bases that are orthogonal to one another. By designating these bases to different users in adjacent cellular communication cells, the inter-cell interference can be minimized.

1.2.2.5 Reduced sensitivity to channel effects

The performance of communication systems is influenced by the kind of modulation scheme used. The modulation mode in turn is affected by the set of waveforms used. By cleverly altering the nature and characteristics of the waveforms used, the sensitivity of the communication system to harmful channel effects can be reduced [9].

1.2.2.6 Development of generic, multi-purpose transceivers

Furthermore, a generic and parameterized wavelet-based radio can help simplify the system architecture by doing away with multiple firmware, software, drivers that indirectly contribute to reduced power consumption and improved battery life. The radio can be designed merely by altering the parameters instead of adding/removing hardware components to the transceiver chain.

1.2.2.7 Optimization of power utilization

While there is no explicit relationship between power optimization and waveforms, the nature and characteristics of the waveform can be altered to suit a set of requirements that can indirectly contribute to a more efficient system, resulting in lower requirements of power and energy. These criteria could typically be:

- minimization of ISI, ICI or PAPR;
- greater tolerance and robustness to time/frequency/phase offset errors;
- robustness towards interference from competing sources;
- possibilities for opportunistic communication (e.g. cognitive radio) where unused resources can be cleverly utilized.

1.2.2.8 Reduced complexity of implementation

It has been proven [9] that the complexity of the wavelet systems is by and large less than OFDM systems. A lower complexity also means lower power requirements in the execution of the signal processing algorithms. The implementation of wavelet systems can be simplified even further if fast-wavelet transforms are employed.

1.2.3 Application of wavelets for wireless transmission

The wavelet transform holds promise as a possible analysis scheme for the design of sophisticated digital wireless communication systems, with advantages such as flexibility of the transform, lower sensitivity to channel distortion and interference and better utilization of spectrum. Wavelets have found beneficial applicability in various aspects of wireless communication systems design including channel modelling, design of transceivers, data representation, data compression, source and channel coding, interference mitigation, signal de-noising, energy-efficient networking. Figure 1.3 gives a graphical representation of some of the facets of wireless communications where wavelets hold promise [4].

1.2.4 Wavelet-packet-based multi-carrier modulation (WPM) system

The promise of wavelets for wireless systems design is exemplified in this research work by realizing an orthogonal multi-carrier system based on wavelet packets[3]. Orthogonally multiplexed communication is a modulation format that places independent information-carrying symbols on orthogonal signals. These orthogonal signals are typically equi-spaced sub-carriers that are modulated to occupy different centre frequencies. Hence, this signalling is also referred to as multi-carrier modulation (MCM). In traditional implementations of MCM, such as the orthogonal frequency division multiplexing or OFDM, the sub-carrier waveforms are Fourier bases or complex exponential functions. Recently, the wavelet packet transform has emerged as an important signal-processing tool. The basis functions in wavelet packet representation are obtained from a single

[3] Wavelet packets are generalized form of wavelets and will be dealt with in detail in Chapters 2 and 3.

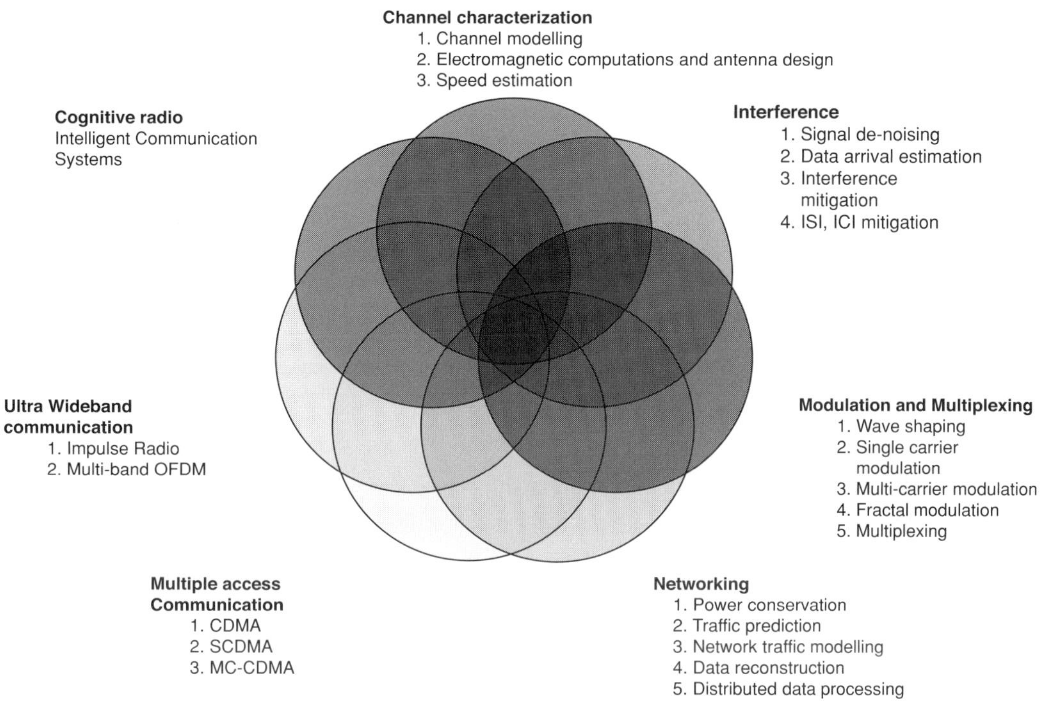

Figure 1.3 The spectrum of wavelet applications for wireless communication.

function called the mother wavelet through scaling and translations. When the scales and translations are dyadic, the resultant basis functions are orthogonal and span[4] embedded sub-spaces of $L^2(\mathbb{R})$,[5] at different resolutions yielding a multi-resolution analysis. From the perspective of communication system design, this has important and interesting implications – finite energy signals in $L^2(\mathbb{R})$ can be decomposed into orthogonal sub-spaces through a wavelet packet transform, or conversely information can be packed into mutually orthogonal wavelet packet basis functions in a way that they do not interfere with one another. Since the basis functions and sub-spaces are orthogonal, such structures can be used in developing orthogonal waveforms for a wavelet-packet-based MCM, leading to the idea of WPM.

The greatest motivation for pursuing WPM in wireless communication systems is in the flexibility and adaptability that they offer [10]–[13]. Unlike OFDM where the carriers are static sine/cosine bases, WPM uses wavelets whose features can be tailored to satisfy an engineering demand. Different wavelets result in different sub-carriers with varying temporal and spectral characteristics and different transmission properties. By careful selection of proper wavelets, it is possible to optimize WPM performance in terms of bandwidth efficiency, frequency selectivity of sub-carriers, sensitivity to synchronization errors, peak-to-average power ratio (PAPR), etc. Furthermore, the WPM

[4] The span of S may be defined as the collection of all (finite) linear combinations of the elements of S.

[5] Set of square-integrable functions in $\mathbb{R}$.

can be efficiently implemented with filter banks obtained by cascading a fundamental quadrature mirror filter pair.

Futuristic wireless transceivers demand a great deal of flexibility and adaptability in order to operate in the crowded spectrum. These requirements correspond to the nature of WPM, making it a strong candidate for the upcoming intelligent communication systems. However, existing knowledge on wavelets for multi-carrier modulation is very limited and the literature on the topic is sparse. Moreover, a lot of key research questions remain to be addressed before WPM can become practically viable. Addressing these issues forms the basis for the efforts of this research book.

1.3 Scope of the book

The objectives stated in the previous section are too broad and ambitious to be covered in a single book. Hence, this research work will confine itself to the mathematical modelling and implementation of the wavelet packet modulator (WPM) on a simulation platform. Furthermore, some of the major technical challenges in the implementation will also be addressed. Only the radio transmission (physical layer) challenges will be considered in the following. The book is organized in eight chapters with the material categorized into three broad divisions, namely, theoretical background (Chapters 1 and 2), wavelet radio (Chapters 3, 4 and 5) and wavelet applications in cognitive radio design (Chapters 6 and 7).

1.3.1 Theoretical background (Chapters 1 and 2)

The theoretical background, is presented over two chapters. The contents provided thus far constitute the first chapter. In Chapter 2, material on the theory of wavelets is provided.

1.3.2 Wavelet radio (Chapters 3, 4 and 5)

In Chapter 3 the wavelet packet modulator, which is the focus of this book, is introduced. We take up three of the issues encountered in the implementation of WPM. Each of these challenges is handled in separate chapters.

In Chapter 4 the influence of loss of synchronization (time/frequency/phase) on the performance of the WPM system is analyzed. For each of these synchronization errors a model is presented and theoretical analysis is given for both WPM and OFDM. The bit error rate (BER) performance under time offset, frequency offset and phase noise is investigated by means of simulation studies. The simulations are performed for WPM with different types of standard wavelets and compared to OFDM.

In Chapter 5 the sensitivity of WPM to PAPR is explored.

1.3.3 Wavelet applications in cognitive radio design (Chapters 6 and 7)

In the final two chapters some of the benefits of pursuing wavelet-based systems for wireless systems design are demonstrated with some examples:

In Chapter 6 a spectrum estimator based on wavelet packets for cognitive radio applications is explained. The proposed method is shown to be efficient in estimation of spectrum and the performances comparable with existing techniques. An efficient wavelet packet spectral estimator (WPSE) based on compressed sensing and with reduced number of sensing measurements is also developed.

In Chapter 7, a general, unified approach to design and develop orthogonal wavelet packet bases according to a requirement is presented. To this end, the design criterion and the wavelet constraints are first listed. The problem that is originally non-linear and non-convex in nature is then converted into a tractable convex optimization problem and finally solved using suitable semi-definite programming (SDP) tools. The proposed mechanism is demonstrated through two *toy* examples where families of wavelets that are i) maximally frequency selective and ii) have the lowest cross-correlation energy, respectively, are developed. The design procedure borrows from the studies conducted in earlier chapters. For example, the design of maximally frequency selective filters borrows from the studies of Chapter 6, while the construction of filters with low cross-correlation uses the conclusions of Chapter 4.

Finally, the book rounds off in Chapter 8 with conclusions and recommendations for future research.

References

[1] R.W. Brodersen, A. Wolisz, D. Cabric, and S.M. Mishra, "CORVUS: A Cognitive Radio Approach for Usage of Virtual Unlicensed Spectrum," Berkeley Wireless Research Center (BWRC), White paper, 2004.

[2] Cisco Visual Networking Index: Global Mobile Data Traffic Forecast Update, 2011–2016, Retrieved July 2012, from http://www.cisco.com/en/US/solutions/collateral/ns341/ns525/ns537/ns705/ns827/white_paper_c11–520862.pdf

[3] Fiber To The Home (FTTH) Council, Fiber to the Home Advantages of Optical Access, Retrieved July 2012, from http://www.salisburync.gov/ftth/fiber_advantages.pdf

[4] M.K. Lakshmanan and H. Nikookar, "A Review of Wavelets for Digital Wireless Communication," *Wireless Personal Communications*, vol. 37, Numbers 3–4, May 2006, pp. 387–420(34), Springer.

[5] G. Wornell, "Emerging Applications of Multirate Signal Processing and Wavelets in Digital Communications," *Proceedings of IEEE*, vol. 84, no. 2.2, pp. 586–603, April 1996.

[6] A. Lindsey, "Wavelet Packet Modulation for Orthogonally Transmultiplexed Communications," *IEEE Trans. Signal Processing*, vol. 45 no. 5, pp. 1136–9, 1997.

[7] P. Vaidyanathan, *Multirate Systems and Filter Banks*, Prentice Hall, NJ, 1993.

[8] L. Ramac and P. Varshney, "A Wavelet Domain Diversity Method for Transmission of Images over Wireless Channels," *IEEE J. Select. Area. Communications*, vol. 18, no. 6, June 2000.

[9] A. Jamin and P. Mahonen, "Wavelet Packet Modulation for Wireless Communications," *Wireless Communications and Mobile Computing Journal*, vol. 5, Issue 2, pp. 123–37, John Wiley and Sons Ltd., March 2005.

[10] H. Nikookar, "Wavelet Radio: Smart, Adaptive and Reconfigurable Radio Systems Based on Wavelets," *European Conference Wireless Technology*, October 2007, Munich, Germany.

[11] H. Nikookar, "Wavelets for Wireless Communication," *10th International Symposium on Wireless Personal Multimedia Communications*, December 2007, Jaipur, India.

[12] H. Nikookar, "Wavelet based Multicarrier Communication Techniques for Cognitive Radios," *First International Conference on Wireless VITAE* 2009, Aalborg, Denmark.

[13] H. Nikookar, "An Introduction to Wavelets for Cognitive Radio," *12th International Symposium on Wireless Personal Multimedia Communications (WPMC)*, September 2009, Sendai, Japan.

2 Theory of wavelets

The wavelet transform is a powerful new tool to analyze data. It can be used to represent known/unknown signals as a set of known functions, called wavelets, and gain insights on their characteristics. The tool is used in various applications and is becoming very popular amongst technologists, engineers and mathematicians alike. In most of the applications, the power of the transform comes from the fact that the basis functions of the transform have compact support in time (or space) and are localized in frequency. Furthermore, the technique allows analysis of signals at different resolutions that often correspond to the natural behaviour of the process one wants to understand. These properties make wavelet transform a natural choice in fields as diverse as image synthesis, data compression, computer graphics and animation, human vision, radar, optics, astronomy, acoustics, seismology, nuclear engineering, biomedical engineering, magnetic resonance imaging, music, fractals, turbulence and pure mathematics [1]. Recently, a wavelet transform has also been used in the design of sophisticated digital wireless communication systems including channel modelling, transceiver design, data representation and compression, source/channel coding, interference mitigation, signal de-noising and energy-efficient networking [2].

In this chapter we provide an overview of the mathematical foundations of the wavelet theory. The material provided in this chapter will not only aid the understanding of later chapters but also serve to make the book self-contained. A thorough study of the subject can be found in [1]–[18].

We start the chapter with a discussion on the representation of signals in Section 2.1. In this regard we trace the progression of the field of signal representation from classical Fourier analysis through the Gabor transform to the wavelet transform. The sections that follow Section 2.1 will elaborate further on the theory of wavelets. The two major branches of wavelet transform, namely continuous wavelet transform (CWT) and discrete wavelet transform (DWT), are explained in Section 2.2 and Section 2.4, respectively. Section 2.3 will detail an important facet of the wavelet theory known as multi-resolution analysis (MRA). This will be followed by a discussion on the filter bank implementation of DWT that includes material on analysis and synthesis of signals using filter banks in Section 2.5. An important variant of the wavelet transform known as a wavelet packet transform will be presented in Section 2.6. Finally, a review on a few popular wavelet families is given in Section 2.7.

Notations

Throughout this book, continuous variables are enclosed in curved brackets, e.g., $f(x)$, $g(t)$, while discrete variables are denoted in square brackets, e.g., $f[n]$, $g[k]$. Vectors are denoted in boldface, e.g., $\mathbf{z} = z[n] = [z_0\ z_1\ z_2 \ldots\ z_{N-1}]$. The discrete index for time is represented by n, while t is used to connote the continuous time variable. The corresponding indices in the frequency domain are denoted with f (continuous) and k (discrete). Finally, most variables in time/space domain are given in small cases, while their representation in transform domain (Fourier/Gabor/Wavelet) is expressed in upper case.

2.1 Introduction

2.1.1 Representation of signals

Mathematical representation of signals or transforms are a way to describe information or data of a physical signal in terms of known mathematical functions. Through transformations valuable insights on the signal can be gained that can be exploited for various practical purposes. Burke [8] considers the transforms to be *mathematical prisms* that facilitate a better interpretation of signals in just the same way optical prisms split light into colours to enable a better understanding of light. The applications can be as diverse as processing audio/video/image data to modelling geological processes such as tsunamis or earthquakes.

A mathematical transform is usually a linear expression where any given signal $f(x)$ in space $\mathbb{S}$ is expressed as a linear combination of a set of known signals $\{\varphi_i\}_{i\in\mathbb{Z}}$ as [5]:

$$f(x) = \sum_i \alpha_i \varphi_i \tag{2.1}$$

Here, α_i are the expansion coefficients or weights that tell how much of the component φ_i is available in the original signal $f(x)$. The space $\mathbb{S}$ can be finite-dimensional like the set of all real numbers $\mathbb{R}^n$ or the set of all real integers $\mathbb{Z}^n$ or infinite-dimensional like the set of all square integrable functions L^2 or the set of square sum able functions l^2.

The set is said to be complete for the space if there exists a dual set $\{\tilde{\varphi}_i\}_{i\in\mathbb{Z}}$ such that the expansion coefficients α_i can be computed from them; i.e.,

$$\alpha_i = < f(x), \tilde{\varphi}_i > \tag{2.2}$$

Here $<, >$ represents an inner product operation. The set $\{\varphi_i\}_{i\in\mathbb{Z}}$ is considered to be orthonormal and complete when $\varphi_i = \tilde{\varphi}_i$ and

$$< \varphi_i, \varphi_j > = \delta[i - j] \tag{2.3}$$

Here, $\delta[\cdot]$ is the Dirac delta function. On the other hand, the set is said to be biorthogonal if it is complete and the vectors φ_i are linearly independent (but not orthonormal) and satisfy the relation:

$$< \varphi_i, \tilde{\varphi}_j > = \delta[i - j] \tag{2.4}$$

The choice on the right set of basis functions depends on the type of signal to be represented and the application in hand.

2.1.2 Fourier analysis

The earliest recorded work on signal representation was conducted by Jean Baptiste Joseph Fourier in the early nineteenth century. He investigated problems of diffusion of heat and proved that periodic functions can be represented as a series of harmonically related sinusoids. This work, popularly known as the Fourier series expansion, was published in the Théorie Analytique de la Chaleur (The Analytical Theory of Heat) in the year 1822 [8]. While Fourier series allowed representation of periodic functions, a variant called a Fourier transform extended the field to enable decomposition of non-periodic functions with finite energy. It is an integral transform that expresses any complex-valued function of a real variable $x(t)$ in terms of trigonometric basis functions:

$$X(f) = \int_{-\infty}^{\infty} x(t)e^{-j2\pi ft}dt \quad f \in \mathbb{R} \tag{2.5}$$

In signal-processing applications $x(t)$ exists in the time (or space) domain and the transform $X(f)$ represents $x(t)$ in the frequency domain. This is analogous to what music composers do when they represent musical chords in terms of the constituent notes.

Through the reverse transform $x(t)$ can be reconstructed from $X(f)$ as follows:

$$x(t) = \int_{-\infty}^{\infty} X(f)e^{j2\pi ft}df \quad t \in \mathbb{R} \tag{2.6}$$

Since the Fourier transform analyzes a time-based signal to provide frequency information, the operation is regarded as frequency–amplitude decomposition.

The nice thing about the Fourier operations is that the frequency information one obtains after the transforms corresponds with the actual physical waves that make the signal [8].

2.1.3 Gabor transform

The Fourier transform offers excellent frequency resolution but fails to provide any information on the temporal variations[1]. Furthermore, the sines/cosines, which form the building blocks of these operations, stretch to infinity in time. It is therefore important to have a representation that gives both time and frequency information of the signal studied. Dennis Gabor[2] adapted the Fourier transform to analyze only a small section of the signal at a time. In his adaptation, called the short-time Fourier transform (STFT), the signal is windowed into small segments (taken to be stationary) which are then studied independently [7]. For a window function $w(t)$ the STFT operation maps a signal

[1] The temporal data after a Fourier transform is not totally lost but encoded as phase information, which is usually inaccessible.

[2] He won the Nobel Prize in 1971 for his investigation and development of holography.

or function $f(t)$ into a two-dimensional function of time τ and frequency f and can be defined as:

$$\text{STFT}\{x(t)\} \equiv X(\tau, f) = \int_t [x(t)w(t-\tau)] \exp(-j2\pi f t) dt \tag{2.7}$$

The STFT is a compromise between time- and frequency-based views of a signal [19]. A trade-off between the time and frequency resolution is enabled in STFT by altering the dimensions of the window function. Smaller windows offer better time resolution but poorer frequency resolution. On the other hand, if the size of the window is enlarged to allow better frequency resolution, the time resolution is compromised. Another drawback is that once a time window is chosen, it remains the same for all frequencies. Many signals require a more flexible approach, one where the window size can be varied to accurately determine both time and frequency. The solution – *wavelet analysis.*

2.1.4 Wavelet analysis

The wavelet transform is a multi-resolution analysis (MRA) mechanism where an input signal is decomposed into different frequency components, and then each component is studied with resolutions matched to its time-scales. The Fourier transform also decomposes signals into elementary waveforms, but these bases are trigonometric functions (sines and cosines). Thus, when one wants to analyze the local properties of the input signal, such as edges or transients, the Fourier transform is not an efficient analysis tool. By contrast, the wavelet transforms that use irregularly shaped wavelets offer a better representation of sharp changes and local features. The wavelet transform gives good time resolution and poor frequency resolution at high frequencies and a good frequency resolution and poor time resolution at low frequencies. This approach is logical when the studied signal has high-frequency components for short durations and low-frequency components for long durations. Fortunately, the signals that are encountered in most applications are often of this type.

The theory of wavelets emerged from multiple backgrounds (see Figure 2.1) – as continuous wavelet transform (CWT) in geophysics, as sub-band coding in speech and image processing, as filter banks from the fields of signal processing and audio compression, as multi-resolution analysis from computer vision, as pyramid coding from image coding and as atomic decompositions in applied mathematics. These topics had been studied independently under different names by different scientific communities and only recently did these ideas converge to facilitate a unified understanding of the subject. Even though the wavelet nomenclature is diverse, the wavelet theory can be interpreted broadly in terms of its continuous time and discrete time representations. We shall present these topics in the coming sections.

2.2 Continuous wavelet transform

Wavelets were introduced by Mortlet and Grossmann [10], who showed that continuous-time functions $f(t)$ in $L^2(\mathbb{R})$ can be represented by a set of functions $\{\psi_{\kappa,\chi}(t)\}$ obtained

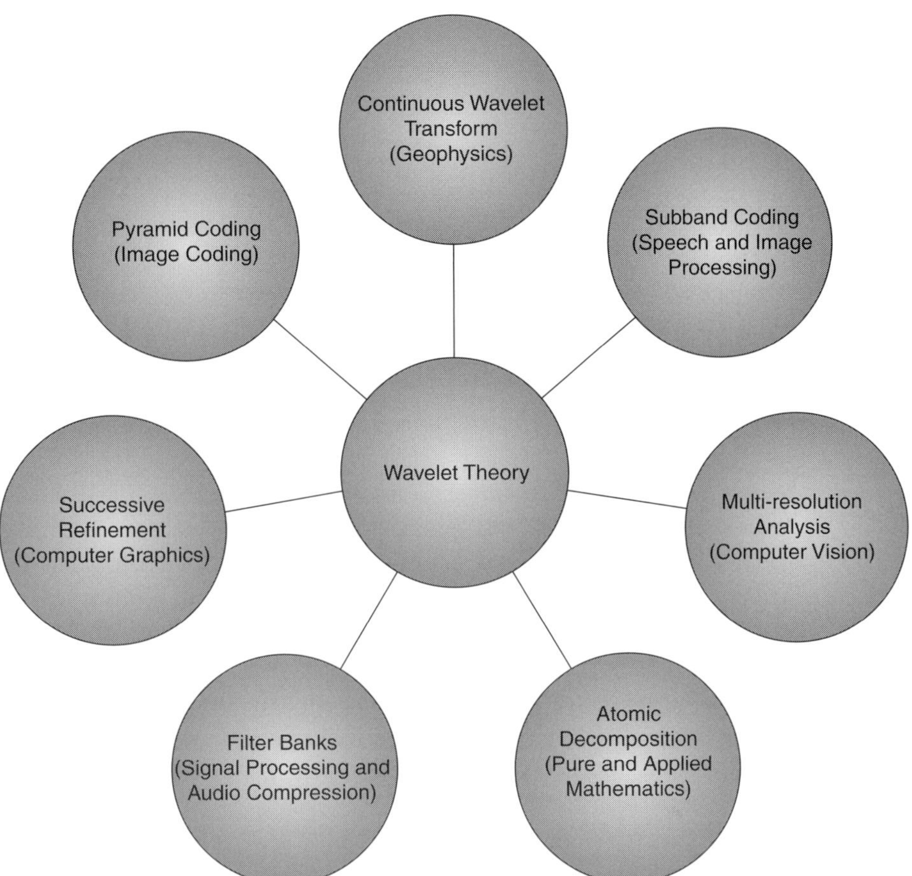

Figure 2.1 Wavelet nomenclature: The figure depicts various terms associated with wavelet theory and the respective domains, enclosed within close brackets, from which the terminologies originated.

by scaling κ and shifting χ primary functions known as mother wavelets $\psi(t)$. The continuous wavelet transform (CWT) of any continuous square-integrable function or signal $f(t)$ in terms of wavelets $\{\psi_{\kappa,\chi}(t)\}$ can be expressed as [6]:

$$\gamma_{\kappa,\chi} = \int_{-\infty}^{\infty} f(t)\psi^*_{\kappa,\chi}(t)dt, \quad \kappa \in \mathbb{R}^+, \chi \in \mathbb{R} \tag{2.8}$$

The expression (2.8) is a general form of CWT, where $\gamma_{\kappa,\chi}$ give the wavelet coefficients of the continuous signal $f(t)$ as a function of the various scaled κ and the shifted χ versions of the mother wavelet $\psi(t)$. The operator * stands for complex conjugation. The mother wavelet $\psi(t)$ is continuous in both time and frequency and the set of baby (or daughter) wavelets functions $\psi_{\kappa,\chi}(t)$ are obtained by scaling κ and shifting χ the mother wavelet [6]:

$$\psi_{\kappa,\chi} = \frac{1}{\sqrt{\kappa}}\psi\left(\frac{t-\chi}{\kappa}\right), \quad \forall\kappa \in \mathbb{R}^+, \chi \in \mathbb{R} \tag{2.9}$$

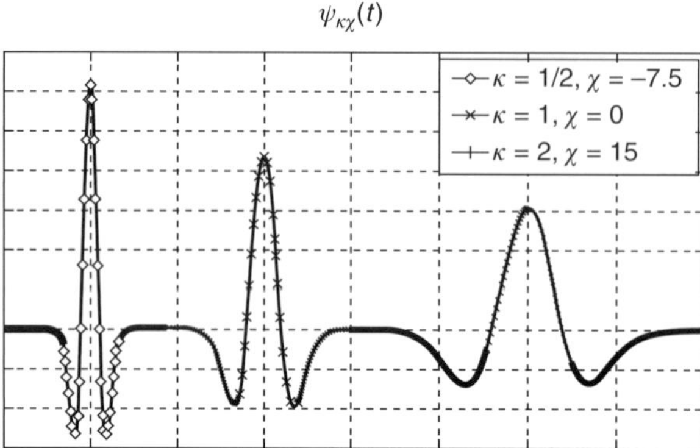

Figure 2.2 Mexican hat wavelet at different translations and scales.

The scaling parameter is similar to the frequency variable in a Fourier transform. It describes how a wavelet basis function is stretched or contracted. On the other hand, the shift variable, also known as the translation parameter, represents the location of the wavelet in time. Both these parameters are continuous-real variables. An example of scaled and translated wavelet is illustrated in Figure 2.2, where a wavelet, popularly known as the Mexican hat because of its shape, is shown for three different translation and scale factors. The wavelet shown at the origin represents the mother wavelet, which is neither shifted nor scaled.

The original signal $f(t)$ can be reconstructed from wavelet coefficients through the inverse wavelet transform [20]:

$$f(t) = \frac{1}{C_\psi} \int_\kappa \int_\chi \gamma_{\kappa,\chi} \tilde{\psi}\left(\frac{t-\chi}{\kappa}\right) \frac{d\chi d\kappa}{\kappa^2} \tag{2.10}$$

where $\tilde{\psi}(t)$ is the dual function of $\psi(t)$ and must satisfy the condition [20]

$$\int_0^\infty \int_{-\infty}^\infty \psi\left(\frac{t_1-\chi}{\kappa}\right) \tilde{\psi}\left(\frac{t-\chi}{\kappa}\right) \frac{d\chi d\kappa}{|\kappa|^3} = \delta(t-t_1) \tag{2.11}$$

For orthogonal expansion sets, $\tilde{\psi}(t) = C_\psi^{-1}\psi(t)$ where [20]

$$C_\psi = \int_{\mathbb{R}} \frac{|\Psi(\omega)|^2}{|\omega|} d\omega \tag{2.12}$$

In Eq. (2.12) $\Psi(\omega)$ denotes the Fourier transform of $\psi(t)$.

An example of the CWT where a signal of finite support is expressed as a two-dimensional (2D) and three-dimensional (3D) time-scale array of coefficients is illustrated in Figure 2.3. The signal considered is a fractal developed by the Swedish mathematician Helge von Koch. The large amplitude in the figure corresponds to high frequency-correlation of the signal with the wavelet function of a particular scale at a certain time instance.

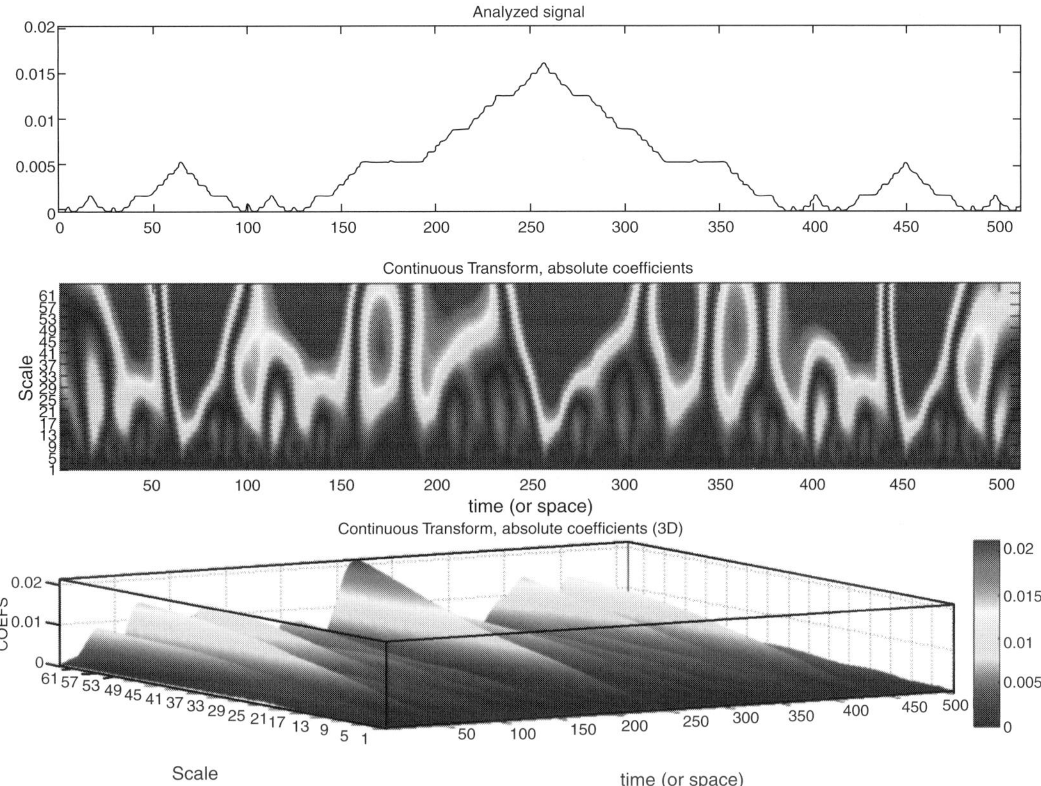

Figure 2.3 Translation-scale representation of a signal.

2.2.1 Orthonormal wavelets

In theory, any function that has zero integral can be considered as the mother wavelet $\psi(t)$. Furthermore, the shift and scale parameters can be real-continuous values ($\kappa \in \mathbb{R}^+$, $\chi \in \mathbb{R}^+$). Hence, the CWT, as expressed in Eq. (2.8), leads to a representation that is infinitely redundant in nature. Such an expression is unwieldy and difficult to implement. To get around this problem, a sparse representation that gives perfect reconstruction of the signal while avoiding redundancy is preferred. The answer is orthogonal wavelets.

Meyer [11] proved that there exist wavelets $\psi(t)$ that provide an orthogonal expansion set of $L^2(\mathbb{R})$ and is of the form:

$$\psi_{\alpha,\beta} = \sqrt{2^{\alpha}}\psi(2^{\alpha}t - \beta), \quad \forall \alpha, \beta \in \mathbb{Z} \tag{2.13}$$

In Eq. (2.13) α and β are the scaling and shift parameters that vary in discrete integer units, i.e., $\alpha, \beta \in \mathbb{Z}$. Meyer also showed that these wavelets are generalized form of the Haar function. The work of Meyer was carried forward by Daubechies [4], [21], who produced a family of wavelets that in addition to being orthogonal also had compact support.

2.2.2 Non-dyadic wavelets

It is important to note that orthonormal wavelets need not always be of the form (2.13), nor do the scales have to be dyadic. In fact, recent studies show that the scaling factor can be different from 2 and can take any rational value $p/q > 1$ [21]. However, in these more general cases, it may be necessary to introduce more than one (but always a finite number) of mother wavelets.

Throughout this book only orthonormal wavelets of the form (2.13) will be used. Not only is the theory of dyadic wavelets well established, but the bases with factor 2 are also easy to implement for numerical computations.

2.3 Multi-resolution analysis

An important advancement in the field of wavelets was the multi-resolution analysis (MRA) framework developed by Mallat [22] and Meyer [23]. The MRA allows characterization of $\psi(t) \in L^2(\mathbb{R})$ that result in an orthonormal basis. The starting point in the discussion on MRA is to consider the wavelet coefficients $\langle f(t), \psi_{\alpha,\beta}(t)\rangle$ at any scale α, which covers the difference in the approximations of $f(t)$ at resolutions $2^{\alpha+1}$ and 2^{α}, respectively. To characterize the successive vector spaces V_α in which the function $f(t)$ is approximated, a complementary function called the scaling function $\varphi(t)$ is defined[3]. As in the case of wavelet functions $\psi_{\alpha,\beta}(t)$, there also exists an extended family of scaling functions $\varphi_{\alpha,\beta}(t)$ that are obtained by a time-shifted version of the fundamental scaling function $\varphi(t)$ [9]:

$$\varphi_{\alpha,\beta} = \sqrt{2^\alpha}\varphi(2^\alpha t - \beta), \quad \forall\alpha, \beta \in \mathbb{Z}, \varphi \in L^2 \tag{2.14}$$

The approximation sub-spaces V_α spanned by the scaling functions $\varphi_{\alpha,\beta}(t)$ over integers $-\infty < \beta < \infty$ are defined by:

$$V_\alpha = \overline{\underset{\beta}{\text{Span}}\{\varphi_\beta(2^\alpha t)\}} = \overline{\underset{\beta}{\text{Span}}\{\varphi_{\alpha,\beta}(t)\}} \tag{2.15}$$

Low values of α provide coarse representation of a signal, while higher values of α represent the finer details. MRA requires the spaces V_α spanned by the scaling functions $\varphi_{\alpha,\beta}(t)$ to have finite energy and ordered as a nested approximation space as [9]:

$$\{0\}\ldots \subset V_{-2} \subset V_{-1} \subset V_0 \subset V_1 \subset V_2 \subset \quad \ldots L^2(\mathbb{R})$$

i.e.:

$$\begin{aligned} &V_\alpha \subset V_{\alpha+1} \quad \forall\alpha \in \mathbb{Z} \\ &\cap_{\alpha\in\mathbb{Z}} V_\alpha = \{0\} \\ &\cup_{\alpha\in\mathbb{Z}} V_\alpha = L^2(\mathbb{R}) \end{aligned} \tag{2.16}$$

[3] The scaling functions are also called father wavelets. The father wavelet acts with the mother wavelet to yield a family of baby wavelets.

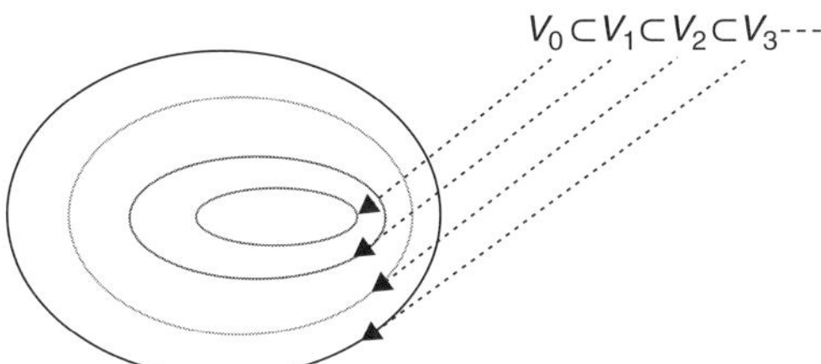

Figure 2.4 Spaces spanned by the scaling functions.

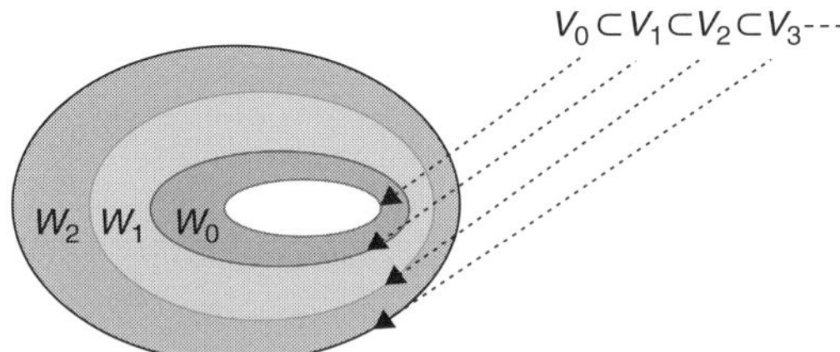

Figure 2.5 Spaces spanned by the scaling functions and wavelets.

Equation (2.16) implies that the space that contains high-resolution approximates of a signal will also contain information on its lower-resolution representation. The nested vector spaces spanned by the scaling functions are illustrated in Figure 2.4.

The MRA imposes strict restrictions on what the scaling function $\varphi(t)$ can be. One of the conditions is that there exists weights/coefficients $h[k]$ such that the scaling function $\varphi(t)$ (which spans V_0) can be expressed as a weighted sum of shifted versions of $\varphi(2t)$ (which spans V_1), i.e. [9]:

$$\varphi(t) = \sum_k h[k]\sqrt{2}\varphi(2t - k), \quad \forall k \in \mathbb{Z} \tag{2.17}$$

There are other restrictions on the nature of $\varphi(t)$ and $\psi(t)$ that are usually determined by the scaling coefficients $h[k]$. We shall delve into them in Chapter 7 where the design of wavelets is discussed in detail.

As mentioned earlier, the wavelets $\psi(t)$ in MRA are defined as orthogonal bases that span the differences between the spaces spanned by the scaling functions at various scales. Let the sub-space spanned by the wavelet be W_{j-1}, and then the function spaces covered by the scaling functions V_α can be written as:

$$\begin{aligned} V_1 &= V_0 \oplus W_0 \\ V_2 &= V_1 \oplus W_1 = (V_0 \oplus W_0) \oplus W_0 \\ &\vdots \\ V_{\alpha+1} &= V_\alpha \oplus W_\alpha = \oplus_{l=0}^{\alpha} W_l, \quad \forall \alpha \in \mathbb{Z} \end{aligned} \tag{2.18}$$

Nested vector spaces spanned by the scaling function and wavelet vector spaces are illustrated in Figure 2.5.

It should be noted that the space W_0 spanned by the wavelets $\psi(t)$ is also a subspace of V_1 ($W_0 \subset V_1$). And therefore, there exists a corresponding orthonormal basis of wavelets defined by a weighted sum of shifted scaling function $\varphi(2t)$,

$$\psi(t) = \sum_k g[k]\sqrt{2}\varphi(2t - k), \quad \forall k \in \mathbb{Z} \tag{2.19}$$

In Eq. (2.19) $g[k]$ denotes the wavelet function coefficients. Because of the orthogonality condition $V_0 \perp W_0 \perp W_1 \perp \cdots \perp W_\alpha$, the scaling and wavelet coefficients are related to each other by [6]–[7]:

$$g[k] = (-1)^k h[L - 1 - k] \tag{2.20}$$

for $h[k]$ of length L.

2.4 Discrete wavelet transform

For practical applications the continuous wavelet transform is not useful and therefore a discrete version of the wavelet transform is preferred. Assuming an orthogonal transform, the forward discrete wavelet transform (DWT) of a discrete signal or function $f(n)$ $\{n = 0, 1, 2 \ldots M - 1\}$ belonging to $l^2(\mathbb{Z})$ is defined as[4]:

$$\begin{aligned} \lambda_{\alpha,\beta} &= \langle f[n], \varphi_{\alpha,\beta}[n] \rangle \\ &= \frac{1}{\sqrt{M}} \sum_n f[n] \varphi_{\alpha,\beta}[n] \\ &= \frac{1}{\sqrt{M}} \sum_n f[n] 2^{\alpha/2} \varphi[2^\alpha n - \beta] \end{aligned} \tag{2.21}$$

$$\begin{aligned} \gamma_{\alpha,\beta} &= \langle f[n], \psi_{\alpha,\beta}[n] \rangle \\ &= \frac{1}{\sqrt{M}} \sum_n f[n] \psi_{\alpha,\beta}[n] \\ &= \frac{1}{\sqrt{M}} \sum_n f[n] 2^{\alpha/2} \psi[2^\alpha n - \beta] \end{aligned} \tag{2.22}$$

Here, ($\lambda_{\alpha,\beta}$ and $\gamma_{\alpha,\beta}$) are the scaling and wavelet transform coefficients and $\frac{1}{\sqrt{M}}$ is the normalization factor. Usually, the value of M is limited by the desired resolution α and is taken to be $M = 2^\alpha$.

The inverse transform to approximate $f[n]$ in terms of the scaling functions $\{\varphi_{\alpha,\beta}[n]\}$ is given as [9]:

$$f[n] = \frac{1}{\sqrt{M}} \left(\sum_{\alpha=-\infty}^{\infty} \sum_{\beta=-\infty}^{\infty} \lambda_{\alpha,\beta} \varphi_{\alpha,\beta}[n] \right) \tag{2.23}$$

[4] For the discrete version the notation of the time unit has been changed from $t \to n$.

This can be rewritten at a desired resolution space V_{α_0} by a series sum of scaling function of sub-space α_0 and wavelet functions of sub-space $\{\alpha = \alpha_0 \rightarrow \infty\}$ as follows [9]:

$$f[n] = \frac{1}{\sqrt{M}} \left(\underbrace{\sum_{\beta=-\infty}^{\infty} \lambda_{\alpha_0,\beta}\varphi_{\alpha_0,\beta}[n]}_{V_{\alpha_0}} + \underbrace{\sum_{\alpha=\alpha_0}^{\infty} \sum_{\beta=-\infty}^{\infty} \gamma_{\alpha,\beta}\psi_{\alpha,\beta}[n]}_{\subset W_\alpha} \right) \tag{2.24}$$

The parameter α_0 in Eq. (2.24) is an integer which sets the coarsest level of approximation of the function $f[n]$, the details of which are filled by its projection onto the wavelet spaces W_α. In terms of the function spaces the resolution Nr at which $f[n]$ is approximated can be given as:

$$V_{Nr} = V_{\alpha_0} + \sum_{\beta=0}^{Nr-1} W_\beta \tag{2.25}$$

2.5 Filter bank representation of DWT

One of the breakthroughs of wavelet transform was the possibility of implementing the DWT algorithm using filter banks. Mallat [15], [22] showed that it is possible to perform DWT decomposition and reconstruction using 2-channel filter banks through a hierarchical algorithm known as the pyramidal algorithm. The algorithm meant that results of wavelet theory could be developed entirely using filter banks. In the next two sections we shall see how this is done.

2.5.1 Analysis filter bank

We start by considering the discrete variant of Eq. (2.17) that expresses the scaling functions $\varphi[n]$ as a series sum of shifted versions $\varphi[2n]$ [9],

$$\varphi[n] = \sum_k h[k]\sqrt{2}\varphi(2n-k), \quad \forall k \in \mathbb{Z} \tag{2.26}$$

Applying the transform $n \rightarrow 2^\alpha n - \beta$ we obtain,

$$\begin{aligned} \varphi[2^\alpha n - \beta] &= \sum_k h[k]\sqrt{2}\varphi[2(2^\alpha n - \beta) - k] \\ &= \sum_k h[k]\sqrt{2}\varphi[2^{\alpha+1}n - 2\beta - k] \\ &= \sum_{m=2\beta+k} h[m-2\beta]\sqrt{2}\varphi[2^{\alpha+1}n - m] \end{aligned} \tag{2.27}$$

Similarly, considering the discrete version of Eq. (2.19)

$$\psi[n] = \sum_k g[k]\sqrt{2}\varphi\,(2n-k), \forall k \in \mathbb{Z} \tag{2.28}$$

and applying the transform $n \to 2^\alpha n - \beta$ we obtain,

$$
\begin{aligned}
\psi[2^\alpha n - \beta] &= \sum_k g[k]\sqrt{2}\varphi[2(2^\alpha n - \beta) - k] \\
&= \sum_k g[k]\sqrt{2}\varphi[2^{\alpha+1} n - 2\beta - k] \\
&= \sum_{m=2\beta+k} g[m - 2\beta]\sqrt{2}\varphi[2^{\alpha+1} n - m]
\end{aligned}
\tag{2.29}
$$

The DWT coefficients at scale α by coefficients at the higher scale $\alpha + 1$ can be as follows:

$$
\begin{aligned}
\lambda_{\alpha,\beta} = \langle f[n], \varphi_{\alpha,\beta}[n] \rangle &= \frac{1}{\sqrt{M}} \sum_n f[n]\varphi_{\alpha,\beta}[n] \\
&= \frac{1}{\sqrt{M}} \sum_n f[n] 2^{\alpha/2} \varphi[2^\alpha n - \beta]
\end{aligned}
\tag{2.30}
$$

Substituting Eq. (2.27) into Eq. (2.30) we get,

$$
\begin{aligned}
\lambda_{\alpha,\beta} &= \frac{1}{\sqrt{M}} \sum_n f[n] 2^{\alpha/2} \sum_{m=2\beta+k} h[m - 2\beta]\sqrt{2}\varphi[2^{\alpha+1} n - m] \\
&= \frac{1}{\sqrt{M}} \sum_n h[m - 2\beta] \sum_{m=2\beta+k} f[n] 2^{\alpha+1/2} \varphi[2^{\alpha+1} n - m] \\
&= \frac{1}{\sqrt{M}} \sum_n h[m - 2\beta] \lambda_{\alpha+1,\beta}
\end{aligned}
\tag{2.31}
$$

Similarly, we find

$$
\begin{aligned}
\gamma_{\alpha,\beta} = \langle f[n], \psi_{\alpha,\beta}[n] \rangle &= \frac{1}{\sqrt{M}} \sum_n f[n]\psi_{\alpha,\beta}[n] \\
&= \frac{1}{\sqrt{M}} \sum_n f[n] 2^{\alpha/2} \psi[2^\alpha n - \beta]
\end{aligned}
\tag{2.32}
$$

Substituting Eq. (2.29) into Eq. (2.32) yields

$$
\begin{aligned}
\gamma_{\alpha,\beta} &= \frac{1}{\sqrt{M}} \sum_n f[n] 2^{\alpha/2} \sum_{m=2\beta+k} g[m - 2\beta]\sqrt{2}\psi[2^{\alpha+1} n - m] \\
&= \frac{1}{\sqrt{M}} \sum_n g[m - 2\beta] \sum_{m=2\beta+k} f[n] 2^{\alpha+1/2} \psi[2^{\alpha+1} n - m] \\
&= \frac{1}{\sqrt{M}} \sum_n g[m - 2\beta] \gamma_{\alpha+1,\beta}
\end{aligned}
\tag{2.33}
$$

Equations (2.31) and (2.33) imply that wavelet and scaling DWT coefficients at a certain scale can be calculated by taking a weighted sum of DWT coefficients from higher scales. This can be viewed as convolution between the DWT coefficients at scale $\alpha + 1$ with wavelet and scaling filter coefficients and subsequently subsampling each output with factor 2 to obtain new wavelet and scaling DWT coefficients at scale α. Therefore,

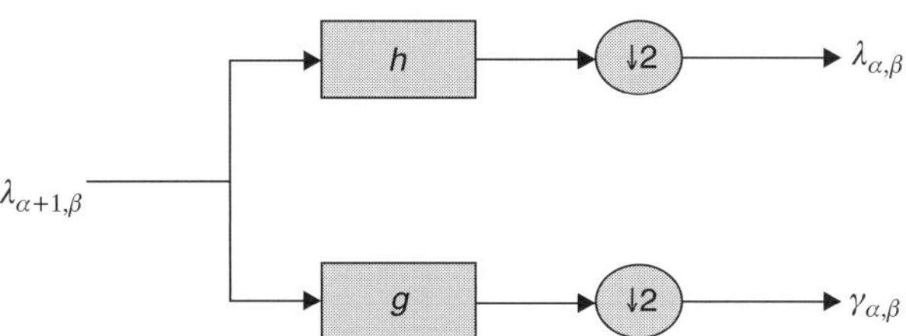

Figure 2.6 2-Channel analysis filter bank.

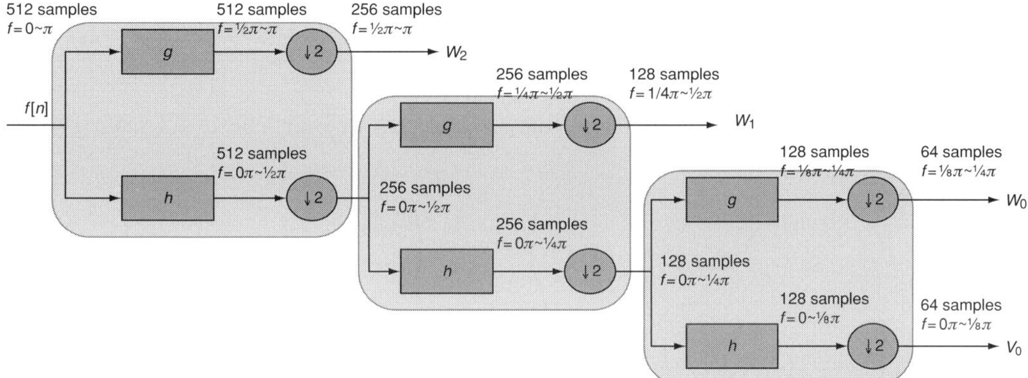

Figure 2.7 3-Stage analysis tree.

we can describe Eqs. (2.26) and (2.28) by a 2-channel filter bank as illustrated in Figure 2.6.

The 2-channel filter bank first splits the input signal in two parts and filters one part with filter h and other with filter g. Both filtered signals are then sub-sampled by 2 and the resulting signals are forwarded to the output of the 2-channel filter bank. Each output signal will therefore contain half the number of samples and will span half of the frequency band compared to the input signal. It should be noticed that the number of samples at the input of the filter bank equals the number of samples at the output.

The complete representation of the DWT can be obtained by iteration of the 2-channel filter bank and taking repeatedly scaling DWT coefficients λ as input. The number of stages in the iteration process will determine the DWT resolution and therefore the number of channels.

The example of a two-band analysis tree with three stages is graphically shown by Figure 2.7. The input signal f has 512 samples and contains frequencies that lie between 0 and π. The resulting decompositions together will still contain 512 samples and span the same frequency band as the original signal but these will be decomposed in different DWT coefficients.

The sub-band structure of wavelet decomposition in the frequency domain can be calculated using Fourier transformation. For the previous example of a 3-stage analysis tree, the corresponding sub-band structure is illustrated in Figure 2.8.

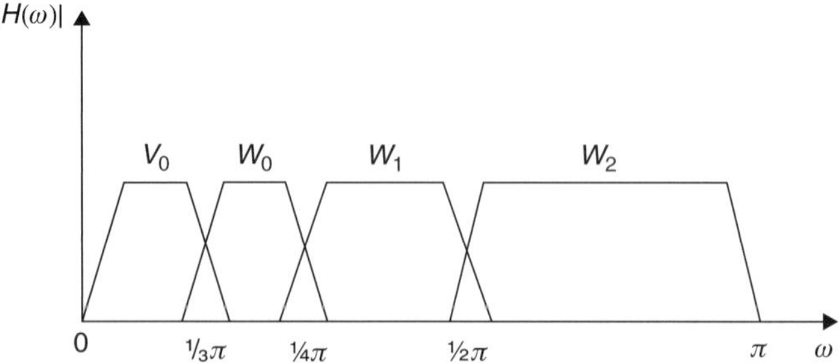

Figure 2.8 Frequency bands for the 3-stage analysis tree.

2.5.2 Synthesis filter bank

The reconstruction formula is derived by considering a signal in the $\alpha + 1$ scaling space $f[n] \in V_{\alpha+1}$ as [9]:

$$f[n] = \frac{1}{\sqrt{M}} \left(\sum_{\beta=-\infty}^{\infty} \lambda_{\alpha+1,\beta} \varphi_{\alpha+1,\beta}[n] \right) = \frac{1}{\sqrt{M}} \left(\sum_{\beta=-\infty}^{\infty} \lambda_{\alpha+1,\beta} \sqrt{2^{\alpha+1}} \psi[2^{\alpha+1} n - \beta] \right) \tag{2.34}$$

This can be expressed in terms of the next scale as [9]:

$$f[n] = \frac{1}{\sqrt{M}} \left(\sum_{\beta} \lambda_{\alpha,\beta} 2^{\alpha/2} \varphi[2^{\alpha} n - \beta] + \sum_{\beta} \gamma_{\alpha,\beta} 2^{\alpha/2} \psi[2^{\alpha} n - \beta] \right) \tag{2.35}$$

Substituting the 2-scale Eqs. (2.26) and (2.28) into (2.35), we get

$$f[n] = \frac{1}{\sqrt{M}} \left(\sum_{\beta} \lambda_{\alpha,\beta} \sum_{m=2\beta+k} h[m - 2\beta] 2^{(\alpha+1)/2} \varphi[2^{\alpha+1} n - m] \right.$$
$$\left. + \sum_{\beta} \gamma_{\alpha,\beta} \sum_{m=2\beta+k} g[m - 2\beta] 2^{(\alpha+1)/2} \varphi[2^{\alpha+1} n - m] \right) \tag{2.36}$$

Multiplying both sides of Eq. (2.36) by $\varphi[2^{\alpha+1} n - \beta']$ and taking the summation allows us to describe the DWT coefficients at higher scales by those of the lower scale [9]:

$$\lambda_{\alpha+1,\beta} = \sum_{m} \lambda_{\alpha,\beta} h[\beta - 2m] + \sum_{m} \gamma_{\alpha,\beta} g[\beta - 2m] \tag{2.37}$$

The expression (2.37) implies that the DWT coefficients at a certain scale level $\alpha + 1$ can be reconstructed by taking a combination of weighted wavelet and scaling DWT coefficients at the previous scale α.

Introducing two new variables $\hat{h}[n]$ and $\hat{g}[n]$ that are time-reversed versions of $h[n]$ and $g[n]$, i.e., $\hat{h}[n] = h[-n]$ and $\hat{g}[n] = g[-n]$, the expression (2.37) can be described by the 2-channel synthesis filter bank, illustrated in Figure 2.9.

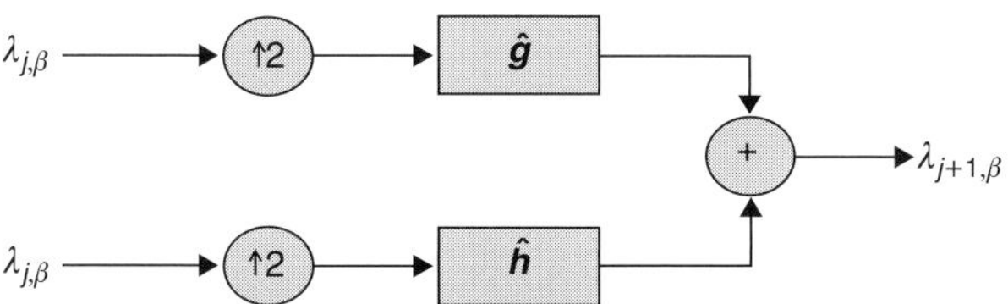

Figure 2.9 2-Channel synthesis filter bank.

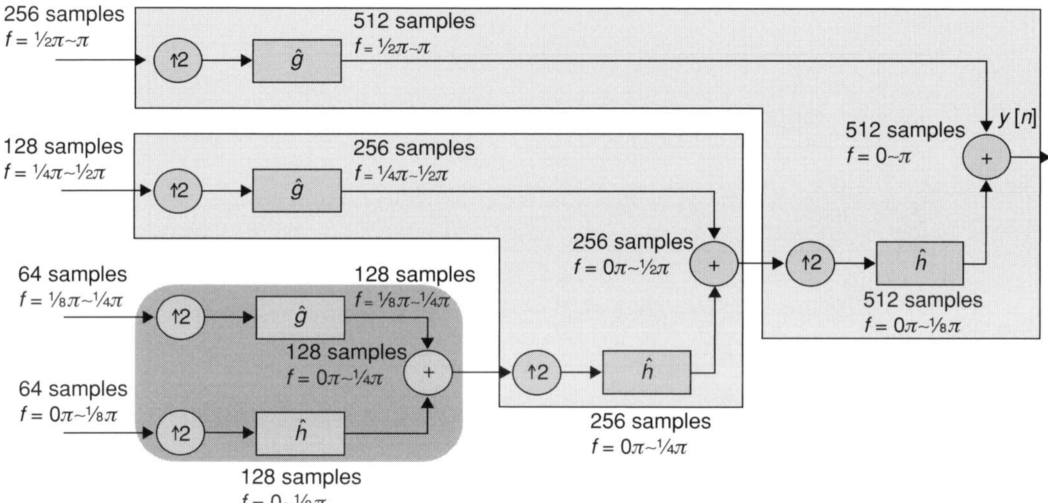

Figure 2.10 Synthesis tree.

The 2-channel synthesis filter bank performs operations that are exactly opposite to those of the analysis filter bank discussed in the previous section. The wavelet and scaling DWT coefficients are first upsampled by a factor of 2, and then the wavelet function DWT coefficients are filtered with HPF $\hat{g}$ while scaling function DWT coefficients are filtered with LPF $\hat{h}$. The two filtered signals are then added to each other to construct DWT coefficients at a higher scale.

The decomposition of a signal in terms of coefficients is called a discrete wavelet transform. In order to reconstruct the original signal from coefficients we can apply the inverse wavelet transform, abbreviated IDWT. The IDWT can be efficiently implemented by iterating the 2-channel synthesis filter bank in the same manner as we have done in the previous paragraph for the 2-channel analysis filter bank. The example of a 3-stage synthesis tree is illustrated in Figure 2.10.

If the assumption of orthogonality is valid, the reconstructed signal is merely a (over Δ) delayed version of the input signal ($x[n] = y[n - \Delta]$). The filter banks that satisfy this property are called perfect reconstruction filter banks.

Figure 2.11 depicts the decomposition of a noisy Doppler function into DWT coefficients at different scales. In this figure we can see how the time-varying frequency signal is described by the wavelet transform as a function of scale and translation

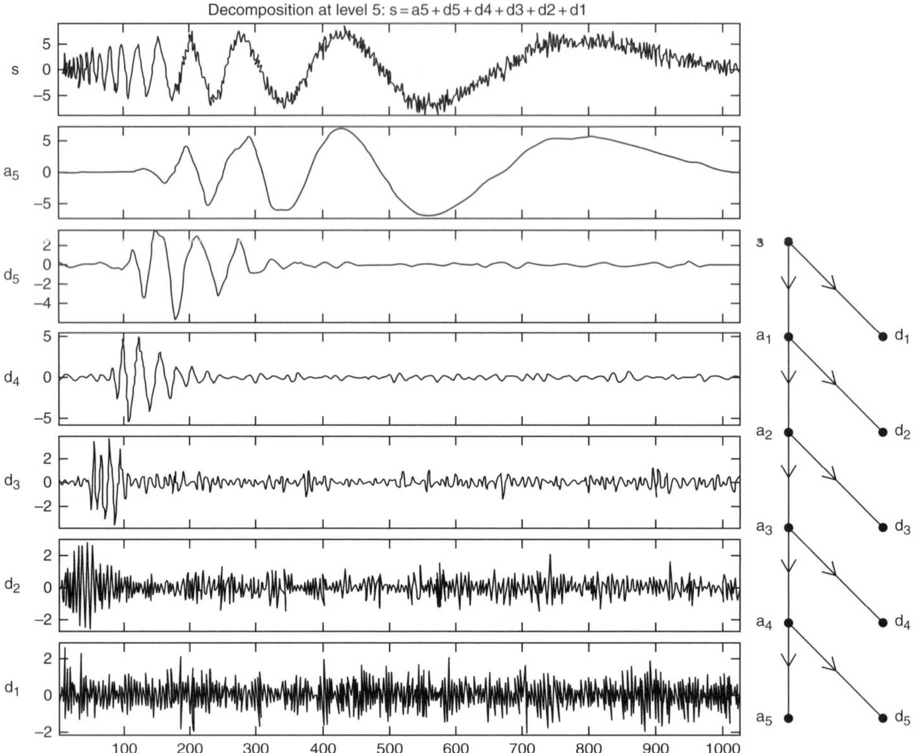

Figure 2.11 Discrete wavelet transform of the noisy Doppler (time domain).

index. A more common and compact figure of DWT performed on the same signal is shown in Figure 2.12. In this representation the depicted colours contain the scale information.

2.6 Wavelet packet transform

The wavelet transform is implemented as a non-uniform filter bank where only the low pass (scaling) branches are iteratively decomposed. The wavelet packet transform is a generalized form of the wavelet transforms where the tree structure used to implement the wavelet algorithm is decomposed on the high pass (wavelet) as well as the low-pass filter branches. The original investigation on the topic was carried out by Coifman and Meyer [24]. And it was followed by Wickerhauser and coworkers [25]–[26], who constructed uniform wavelet packet trees and demonstrated its operation for acoustic signal compression. Because the high frequencies are decomposed in the same manner as low frequencies, the wavelet packet transform has evenly spaced frequency resolution. Figure 2.13 shows the frequency bands of a 3-stage wavelet packet tree.

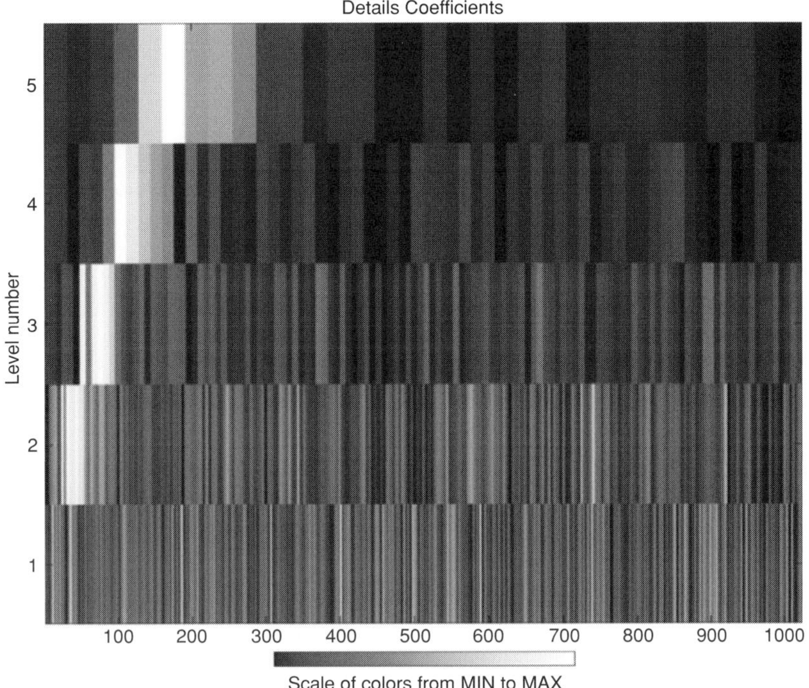

Figure 2.12 Discrete wavelet transform of the noisy Doppler (time-scale domain).

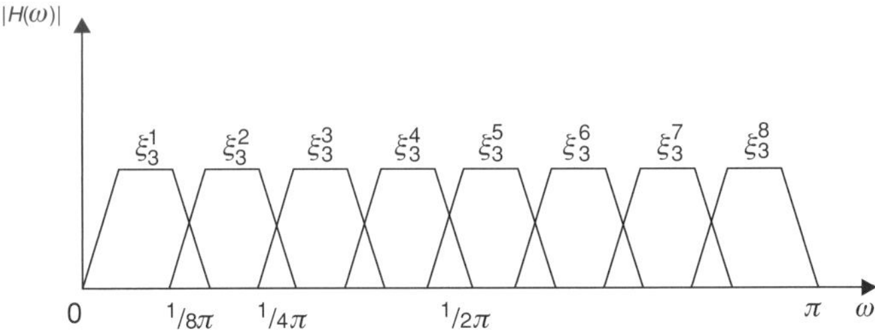

Figure 2.13 Frequency bands for 3-stage wavelet packets tree.

The filter bank structure for a wavelet packet transform usually expands to a full binary tree[5]. In order to make a clear distinction between different sets of coefficients, we label each wavelet packet ξ by the level l that corresponds to the depth of the node in the tree and by the current position p of the node at a given level. Wavelet packet decomposition recursively splits each parent node in two orthogonal sub-spaces W_l^p located at the next level [27]:

$$W_l^p = W_{l+1}^{2p} \oplus W_{l+1}^{2p+1} \tag{2.38}$$

[5] Arbitrary pruning of the full binary tree also lead to a basis for square summable spaces $L^2(\mathbb{R})$.

The sub-spaces given in Eq. (2.38) are those spanned by the basis functions of wavelet packets

$$W_l^p = \overline{\text{span}\{2^{l/2}\xi_l^p[2^l n - k]\}} \tag{2.39}$$

Wavelet packet coefficients ξ at a certain level are calculated by convolving the wavelet and scaling filter with wavelet packets coefficients from a previous level. This action is performed repeatedly for all wavelet packets until the full binary tree is obtained for the desired depth. The wavelet packets coefficients $\xi_{l+1}^{2p}[n]$ are generated using the scaling filter and coefficients $\xi_{l+1}^{2p+1}[n]$ that are created using the wavelet filter [27]–[28]:

$$\begin{aligned} \xi_{l+1}^{2p}[n] &= \sqrt{2}\sum_k h[k]\xi_l^p[2n-k] \\ \xi_{l+1}^{2p+1}[n] &= \sqrt{2}\sum_k g[k]\xi_l^p[2n-k] \end{aligned} \tag{2.40}$$

The expression (2.40) shows the recursive equation for wavelet packets generation. In the regular DWT decomposition for each additional level we need only to perform a single iteration of a 2-channel filter bank, while in the wavelet packet transform the number of iterations is exponentially proportional to the number of levels. Therefore, the wavelet packet transform has higher computational complexity when compared to regular DWT. By utilization of the fast filter bank algorithm the wavelet packet transform requires $O(N\log(N))$ operations, similar to FFT, while DWT needs only $O(N)$ calculations [29].

Figure 2.14 illustrates the full binary tree for a 3-stage wavelet packet analysis.

The reconstruction of wavelet packets is also performed in an iterative method. For each pair of wavelet packets coefficients at level l of the tree, we can calculate wavelet packets coefficients at the previous level $l-1$ by:

$$\xi_l^p[n] = \sum_k h[k]\xi_{l+1}^{2p}[2n-k] + \sum_k g[k]\xi_{l+1}^{2p}[2n-k] \tag{2.41}$$

Figure 2.15 depicts the 3-stage wavelet packets synthesis tree.

Figure 2.16 portrays the wavelet packet decomposition of the noisy Doppler function at different scales. The same noisy Doppler signal as used in the DWT example has also been used here.

2.7 Wavelet types

The wavelet transform is a generic tool with infinitely many wavelets. The nature of the wavelet is entirely determined by the filters that characterize it. Each wavelet has certain distinguishing characteristics that make them more suitable for one application than another. Therefore, during the design of a system careful considerations of the different wavelet properties should be made according to the system requirements.

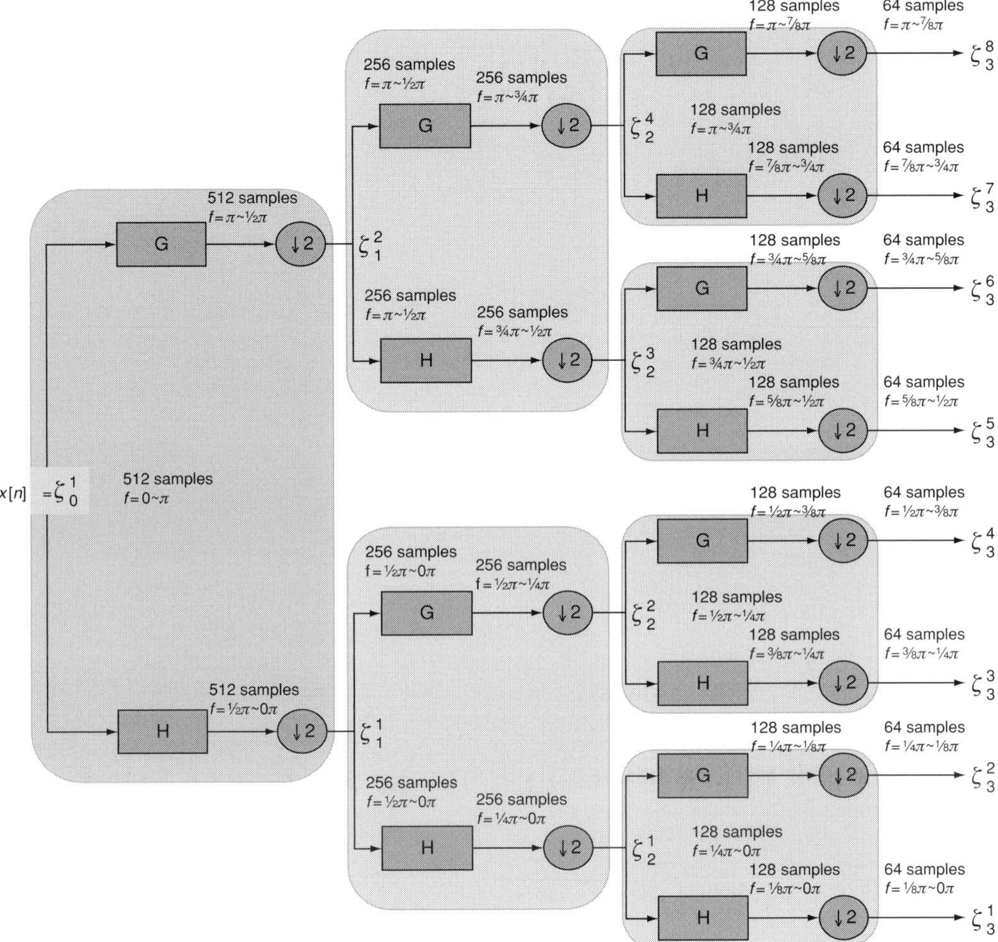

Figure 2.14 3-Stage wavelet packet analysis tree.

2.7.1 Wavelet properties

Many considerations go into the design of a wavelet system including properties such as orthogonality, compact support, symmetry and smoothness. Here, we shall discuss a few important ones.

2.7.1.1 Compact support

This property ensures that the wavelet has a finite number of non-vanishing coefficients and that the filter banks used to derive the wavelets are of finite length [9]. Compact support is defined by the length of the filter. In order to keep the computational complexity to the minimum, usually shorter filters are preferred. However, a longer filter gives more freedom to fine-tune other wavelet properties like orthogonality or regularity.

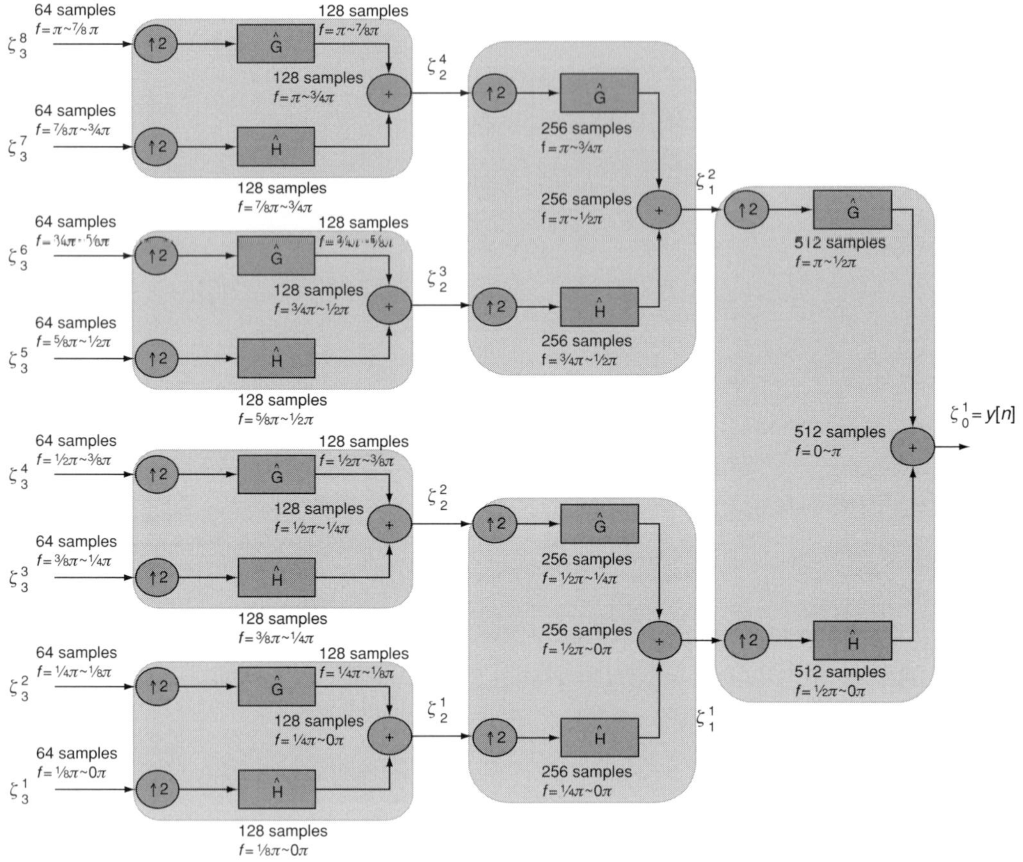

64 samples
$f=\pi\sim\frac{7}{8}\pi$
ζ_3^8
128 samples
$f=\pi\sim\frac{7}{8}\pi$
$\uparrow 2$ $\hat{G}$
128 samples
$f=\pi\sim\frac{3}{4}\pi$
ζ_2^4
64 samples
$f=\frac{7}{8}\pi\sim\frac{3}{4}\pi$
ζ_3^7
$\uparrow 2$ $\hat{H}$
128 samples
$f=\frac{7}{8}\pi\sim\frac{3}{4}\pi$
64 samples
$f=\frac{3}{4}\pi\sim\frac{5}{8}\pi$
ζ_3^6
128 samples
$f=\frac{3}{4}\pi\sim\frac{5}{8}\pi$
128 samples
$f=\frac{3}{4}\pi\sim\frac{1}{2}\pi$
ζ_2^3
64 samples
$f=\frac{5}{8}\pi\sim\frac{1}{2}\pi$
ζ_3^5
128 samples
$f=\frac{5}{8}\pi\sim\frac{1}{2}\pi$
256 samples
$f=\pi\sim\frac{3}{4}\pi$
256 samples
$f=\pi\sim\frac{1}{2}\pi$
ζ_1^2
256 samples
$f=\frac{3}{4}\pi\sim\frac{1}{2}\pi$
512 samples
$f=\pi\sim\frac{1}{2}\pi$
512 samples
$f=0\sim\pi$
$\zeta_0^1=y[n]$
64 samples
$f=\frac{1}{2}\pi\sim\frac{3}{8}\pi$
ζ_3^4
128 samples
$f=\frac{1}{2}\pi\sim\frac{3}{8}\pi$
128 samples
$f=\frac{1}{2}\pi\sim\frac{1}{4}\pi$
ζ_2^2
64 samples
$f=\frac{3}{8}\pi\sim\frac{1}{4}\pi$
ζ_3^3
128 samples
$f=\frac{3}{8}\pi\sim\frac{1}{4}\pi$
256 samples
$f=\frac{1}{2}\pi\sim\frac{1}{4}\pi$
256 samples
$f=\frac{1}{2}\pi\sim 0\pi$
512 samples
$f=\frac{1}{2}\pi\sim 0\pi$
64 samples
$f=\frac{1}{4}\pi\sim\frac{1}{8}\pi$
ζ_3^2
128 samples
$f=\frac{1}{4}\pi\sim\frac{1}{8}\pi$
128 samples
$f=\frac{1}{4}\pi\sim 0\pi$
ζ_2^1
ζ_1^1
64 samples
$f=\frac{1}{8}\pi\sim 0\pi$
ζ_3^1
256 samples
$f=\frac{1}{4}\pi\sim 0\pi$
128 samples
$f=\frac{1}{8}\pi\sim 0\pi$

Figure 2.15 3-Stage wavelet packets synthesis tree.

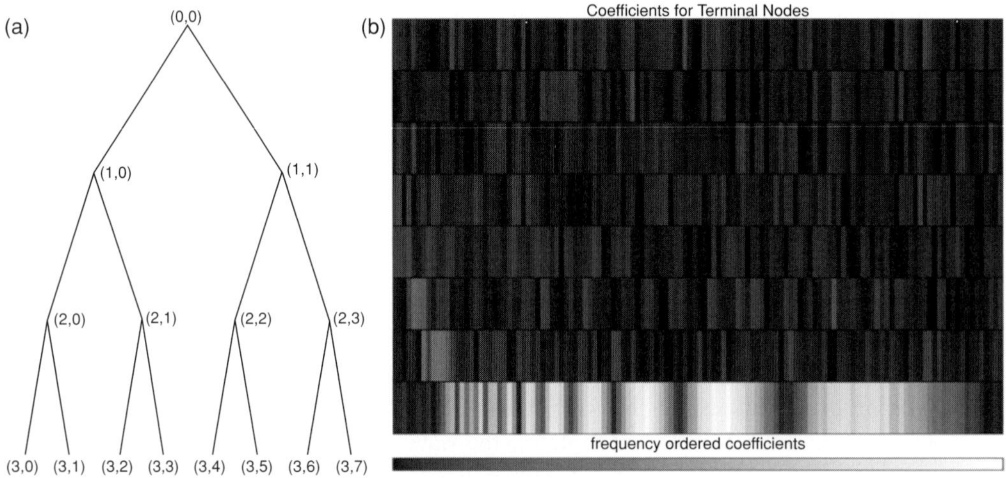

Figure 2.16 Discrete wavelet packet transform of the noisy Doppler. (a) Tree structure, (b) wavelet packet transform in time-scale domain.

2.7.1.2 Paraunitary condition

The paraunitary condition is essential for many reasons. First, it is a pre-requisite for generating orthonormal wavelets [6]–[7]. Secondly, it automatically ensures perfect reconstruction of the decomposed signal [6]; i.e., the original signal can be reconstructed without amplitude or phase or aliasing distortion, if the filter banks used satisfy the paraunitary condition. A rational transfer function $A(z)$ is said to be paraunitary when it obeys the relation $\tilde{A}(z)A(z) = 1$. Here, $\tilde{A}(z)$ is the para-conjugate of $A(z)$ and is given as $\tilde{A}(z) = A^*(z^{-1})$, where the superscript * denotes the conjugation of the coefficients.

Properties 2.7.1.1 and 2.7.1.2 are necessary and sufficient conditions for the wavelets to be realized. However, they may not always guarantee the generation of regular and well-shaped wavelets. Quite often, the wavelets can be irregular or even fractal-shaped. Therefore, to ensure smoothness or regularity of the wavelets the additional property of regularity is important.

2.7.1.3 Regularity

This property is a measure of the smoothness of the wavelet. The regularity condition requires that the wavelet be locally smooth and concentrated in both the time and frequency domains. It is normally quantified by the number of times a wavelet is continuously differentiable. The simplest regularity condition is the *flatness* constraint that is stated on the low-pass filter. An LPF is said to satisfy a Kth-order flatness (or K-regular) if its transfer function $H(z)$ contains K zeroes located at the Nyquist frequency ($z = -1$ or $\omega = \pi$). Parameter K is called the regularity order and for a filter of length L it satisfies the relation $0 \leq K \leq L/2$. K-regularity is also an important measure for wavelets because it helps to reduce the number of non-zero coefficients in the high-pass sub-bands and it is one of the easiest ways to determine if a scaling function is fractal.

Another way to determine the regularity of the wavelets is the number of vanishing moments of the wavelet $\psi(t)$ and scaling functions $\varphi(t)$ [3]. This number is used for the dual vanishing moments to determine the convergence rate of the multi-resolution projections. The jth moments of the wavelet and scaling functions, $m_w(j)$ and $m_s(j)$, respectively, are defined in the continuous-time domain as follows [9]:

$$\begin{aligned} m_w(j) &= \int t^j \psi(t) dt \\ m_s(j) &= \int t^j \varphi(t) dt \end{aligned} \tag{2.42}$$

Usually, the more the contribution from the zero wavelet moments of a wavelet, the smoother will be its scaling function. However, this is not a tight condition. The smoothness is actually defined by the continuous differentiability of the scaling function. There are two ways in which smoothness can be defined: local by the Hölder measure and global by the Sobolev measure. Different measures of smoothness are utilized based on the application in hand. In this book we use the K-regularity as the true measure of smoothness.

2.7.1.4 Symmetry

Symmetrical wavelets have a feature that the transform of the mirror of an image is the same as the mirror of the wavelet transform. None of the orthogonal wavelets except the Haar wavelet is symmetric. Although requiring symmetric wavelets involuntarily means that wavelets are not orthogonal, there are some applications that prefer symmetric wavelets above orthogonal ones. For instance, image-compression techniques like JPEG2000 uses biorthogonal symmetric wavelets. Because by compression of an image we discard one part of the wavelet coefficients containing high detail, the perfect reconstruction has become impossible anyhow. The fulfilment of symmetry property in JPEG2000, on the other hand, results in more natural, smooth images.

2.7.2 Popular wavelet families

A wavelet is defined by the choice of low-pass filter used, obtained after satisfying the compact support, regularity and paraunitary conditions. For a filter of length L this is essentially solving L equations of which $L/2$ come from the paraunitary constraint and K from the regularity/flatness constraint. The remaining $L/2 - K$ conditions offer the freedom to establish a desired wavelet property such as frequency selectivity.

2.7.2.1 Daubechies

The Daubechies are a family of compact supported orthonormal wavelets with the highest degree of smoothness. It was derived by Ingrid Daubechies [4], who used all the degrees of freedom K to generate a wavelet family of maximum regularity for a given filter length L, or minimum L for a given regularity [9]. This she did by imposing the maximum number of zero moments to the wavelet function in the vanishing moments condition.

2.7.2.2 Coiflet

Coiflets are a variation of the Daubechies wavelets. They are so named because it was derived by I. Daubechies at the behest of R. Coifman, who suggested the construction of an orthonormal wavelet basis with vanishing moment conditions for both wavelet and scaling functions (unlike Daubechies where only the wavelet functions have zero moments). The wavelet function has $2L$ moments equal to 0 and the scaling function has $2L - 1$ moments equal to 0.

2.7.2.3 Symlet

The symlet family of wavelets is another variant of the Daubechies family that are *nearly symmetrical* (as opposed to being symmetrical). These modifications were also proposed by I. Daubechies and the properties of the two wavelet families are similar.

In Table 2.1 we list some of the most popular wavelets today and give their most important properties.

Table 2.1 Standard wavelet specifications

Name	Compact Support	Orthogonality	Symmetry	K-Regularity
Haar	2	✔	✔	1
Daubechies	L	✔	✗ far from	$L/2$
Symlets	L	✔	✗ near	$L/2$
Discrete Meyer	102	✔	✗	1
Coiflet	L	✔	✗ near	$L/6$
Bi-orthogonal	(L_1, L_2)	✗	✔	$\approx(L_1/2, L_2/2)$

2.8 Summary

In this chapter we discussed the basics of the theory of the wavelet transform and explained how the discrete wavelet transform can be efficiently implemented with the Mallat's pyramidal tree algorithm using filter banks. Due to the efficient implementation and the freedom they provide, wavelets have emerged in many different fields. Recently, wavelets have been also proposed as a candidate for multi-carrier modulation (MCM). In the next chapter we show how the theory of wavelets and wavelet packets can be applied to the MCM.

References

[1] G. Wornell, "Emerging Applications of Multirate Signal Processing and Wavelets in Digital Communications," *Proc. IEEE*, vol. 84, pp. 586–603, April 1996.

[2] G. Jovanović-Dolecek, *Multirate Systems: Design & Applications*, Hershey, PA: IDEA Group Publishing, 2002.

[3] M.K. Lakshmanan and H. Nikookar, "A Review of Wavelets for Digital Wireless Communication," *Springer Journal on Wireless Personal Communication*, vol. 37, No. 3–4, pp. 387–420, May 2006.

[4] I. Daubechies, *Ten Lectures on Wavelets*, Philadelphia: SIAM, 1992.

[5] G. Strang and T. Nquyen, *Wavelets and Filter Banks*, Wellesley-Cambridge Press, 1996.

[6] M. Vetterli and I. Kovačević, *Wavelets and Subband Coding*, Englewood Cliffs, New Jersey: Prentice Hall PTR, 1995.

[7] P.P. Vaidyanathan, "A Theory for Multiresolution Signal Decomposition: The Wavelet Representation," *IEEE Trans. Pattern Anal. Machine Intell.*, vol. 11, pp. 674–93, July 1989.

[8] B. Burke, *The World According to Wavelets: The Story of a Mathematical Technique in the Making*, Upper Saddle River, New Jersey, A K Peters, May 1998.

[9] C.S. Burns, R.A. Gopinath, and H. Guo, *Introduction to Wavelets and Wavelet Transform, a Primer*, Upper Saddle River, New Jersey, Prentice Hall, 1998.

[10] A. Grossmann and J. Morlet, "Decomposition of Hardy Functions into Square Integrable Wavelets of Constant Shape," *SIAM J. Math.*, vol. 15, pp. 723–36, 1984.

[11] Y. Meyer, Principe d'incertitude, bases hilbertiennes et algebres d'operateurs. Bourbaki seminar, no. 662, 1985–86.

[12] A. Cohen and J. Kovačević, "Wavelets: The Mathematical Background," in *Proc. IEEE*, vol. 84, no. 4, pp. 514–22, April 1996.

[13] G. Kaiser, *A Friendly Guide to Wavelets*, Boston: Bitkhäuser, 1994.

[14] A. Akansu et al., "Wavelet and Subband Transforms: Fundamentals and Communication Applications Frequency," *IEEE Communications Magazine*, vol. 35, pp. 104–15, December 1997.

[15] S.G. Mallat, "A Theory for Multiresolution Signal Decomposition: The Wavelet Representation," *IEEE Transaction on Pattern Recognition and Machine Intelligence*, vol. 11, pp. 674–93, July 1989.

[16] S.G. Mallat, "Multiresolution Approximation and Wavelet Orthonormal Bases of L2," *Transactions of the American Mathematical Society*, vol. 315, pp. 69–87, 1989.

[17] M. Weeks, *Digital Signal Processing using MATLAB and Wavelets*, Hingham: Infinity Science Press LLC, 2007.

[18] H. Nikookar, "Short Course: Wavelets for Wireless Communication," *European Conference on Wireless Technology (ECWT)*, Munich, Germany, October 9–12, 2007.

[19] Online Encyclopedia – http://en.wikipedia.org/wiki/STFT.

[20] Online Encyclopedia – http://en.wikipedia.org/wiki/Continuous_wavelet_transform.

[21] I. Daubechies, "Orthonormal Bases of Compactly Supported Wavelets, II. Variations on a theme," *SIAM J. Math. Anal.*, 24(2), pp. 499–519, 1993.

[22] S. Mallat, "Multiresolution Approximation and Wavelets," *Trans. Amer. Math. Soc.*, 315(1989), pp. 69–88.

[23] Y. Meyer, Ondelletes function splines, et analysys graduees, Lectures given at the Mathematics Department, University of Torino, 1986.

[24] R.R. Coifman and Y. Meyer, "Orthonormal Wave Packet Bases," preprint, Numerical Algorithms Research Group, Yale University, 1989.

[25] R.R. Coifman, Y. Meyer and V. Wickethauser, "Adapted Waveform Analysis, Wavelet Packets and Applications," preprint, Numerical Algorithms Research Group, Yale University, 1992.

[26] M.V. Wickerhauser, *Adapted Wavelet Analysis from Theory to Software*, A. K. Peters, Wellesley, MA (1994).

[27] A.R. Lindsey, Generalized Orthogonally Multiplexed Communication via Wavelet Packet Basis, Ph.D. dissertation, Faculty of the Russ College of Engineering and Technology, Ohio University, 1995.

[28] A. Lindsay, "Wavelet Packet Modulation for Orthogonally Transmultiplexed Communications," *IEEE Transactions on Communications*, vol. 45, pp. 1336–9, May 1997.

[29] A. Jamin and P. Mahonen, "Wavelet Packet Modulation for Wireless Communications," *Wireless Communications & Mobile Computing Journal*, vol. 5, Issue 2, pp. 123–37, John Wiley and Sons Ltd., March 2005.

3 Wavelet packet modulator

Wavelets, filter banks and multi-resolution analysis, which had been developed independently in the fields of applied mathematics, signal processing and computer vision, respectively, have recently converged to form a single theory. In the previous chapter we saw how the theory of wavelets emerged as a natural extension to traditional signal processing tools like Fourier transforms. In this chapter we shall see how a multi-carrier communication system can be constructed with wavelets and wavelet packets.

Multi-carrier modulation is a method where the data to be transmitted are divided into several parallel data streams or channels, one for each sub-carrier. Multi-carrier modulation possesses several properties that make it an attractive approach for high-speed wireless communication networks. Among these properties is the ability to efficiently access and distribute multiplexed data streams, and a reduced susceptibility to impulsive, as well as to narrowband channel disturbances.

In existing multi-carrier transmission schemes, such as the popular orthogonal frequency division multiplexing (OFDM), information-carrying bits modulate orthogonal trigonometric functions that are then added to obtain a composite signal. These implementations use Fourier transforms and are particularly efficient with regard to bandwidth utilization and simplicity of transceiver design. However, they are not without fault – since the building blocks of OFDM are sine/cosine functions that oscillate to infinity in time, the signals usually have to be truncated resulting in deterioration of performance. Furthermore, the basis functions are static and hence the transmission waveforms cannot be altered according to the demands of the wireless transmission. With an ever-increasing demand for high-quality wireless services, there is a growing interest towards alternative orthogonal basis functions that can yield better performances in relation to OFDM. It is in this context that the mathematical precept of wavelets and wavelet packets holds promise.

In this chapter we explain how the theory of wavelets and filter banks can be used to construct a new multi-carrier modulation called wavelet packet modulator (WPM). The theoretical background presented in this chapter will serve as an important prelude to Chapters 4–7 where we address various issues related to the implementation of WPM. But even before we introduce WPM, we quickly review OFDM and other filter-bank-based multi-carrier systems. This discussion on alternative MCM techniques will aid in the better understanding of the WPM technique.

The contents of the chapter are divided into five sections. Section 3.1 gives an overview of existing modulation techniques currently in use for wireless data transmission.

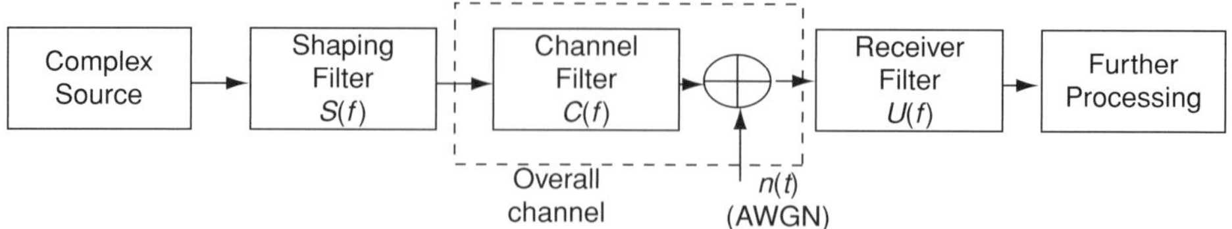

Figure 3.1 Baseband equivalent of a narrowband communication system.

Section 3.2 details the most popular MCM technique, namely, the orthogonal frequency division multiplexing or OFDM. This discussion on OFDM will be followed by an overview of filter-bank-based MCM methods in Section 3.3. Section 3.4 introduces the WPM system implementation. And finally, to round off the chapter a summary of the contents is outlined in Section 3.5.

3.1 Modulation techniques for wireless communication

In telecommunication systems, modulation is a process where information-carrying digital bits are mapped into waveforms (or air waves) so that the message can be physically transmitted. This is done by varying the phase, frequency or amplitude of the waveforms in accordance with the content of the message. While different wireless standards may differ from one another substantially, the air interfaces of all radio platforms operate under one of the three fundamental modulation modes, namely, single-carrier (where the information bits modulate a single waveform or carrier), multi-carrier modulation (where the data is divided into several parallel data streams or channels, one for each sub-carrier) or spread-spectrum (where the signal is transmitted on a bandwidth considerably larger than the frequency content of the original information). Wireless communication systems can hence be viewed as trans-multiplexers characterized by the kind of waveforms they transmit. The properties of the waveform, i.e., the time spread, spectral footprint, shape and the number of carriers, determine the nature of the radio.

3.1.1 Single-carrier transmission

In a single-carrier system the base band signal modulates the carriers using one of the characteristic frequency, phase or amplitude [1]–[2]. Figure 3.1 shows the blocks of a typical narrowband, single-carrier communication system [27]. At the transmitting end, a source generates an arbitrary stream of data derived from the source alphabet. This stream of data is then linearly modulated by a pulse-shaping filter $S(f)$ and then transmitted to the channel. At the receiver the received signal is demodulated and decoded by a receiving filter $U(f)$ and after further processing the data are estimated.

For digital signals, the information is in the form of bits or collections of bits called symbols, which are modulated onto the carrier. When higher bandwidths (data rates) are used, the duration of one bit or symbol of information becomes smaller. At the same time the system becomes more susceptible to loss of information from impulse noise,

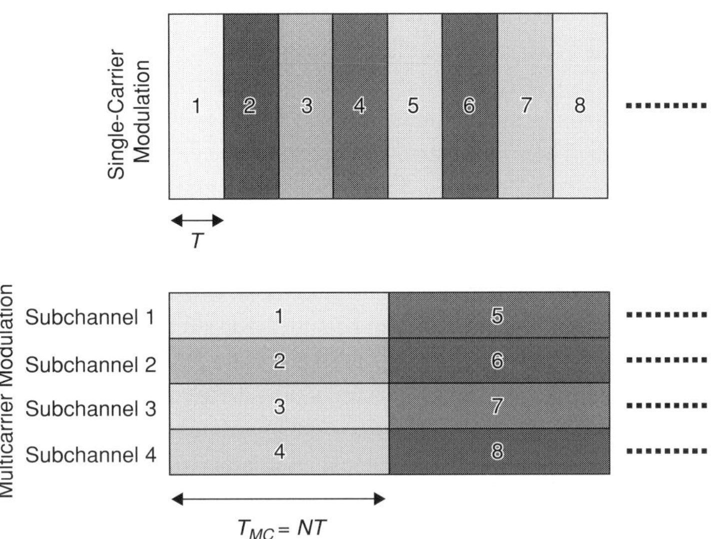

Figure 3.2 Single-carrier and multi-carrier modulation.

signal reflections and other impairments. These impairments can hinder recovery of the transmitted information. In addition, as the bandwidth of the single-carrier system is made larger its vulnerability to channel dispersion is also increased. Therefore, this method is not preferred in practice.

In the past decade the rapid progress of telecommunication market has opened niches for new techniques that can accommodate high data rates without loss in performance. In conventional single-carrier communication systems the data are transmitted sequentially and therefore the duration of each symbol is inversely proportional to the data rate R_s. Higher data rates result in shorter symbol duration. The problem, however, arises in dispersive channels when the duration of transmitted symbols becomes shorter than the delay introduced by the channel. As a result the received symbols are widely spread in time causing inter-symbol interference (ISI). The amount of ISI in a given channel increases with the data rate R_s limiting the connection speed.

ISI can be significantly reduced by employment of multi-carrier modulation (MCM) technique. MCM subdivides the total bandwidth into N narrow channels, which are transmitted in parallel. The original data stream at rate R_s is divided into N streams each having a data rate of $^{R_s}/_N$ and therefore N times longer symbol duration; i.e., $T_{MC} = NT$. Figure 3.2 shows the time–frequency footprints of single and multi-carrier modulated signals.

Multi-carrier modulation (MCM) is the principle of transmitting high data rate by dividing the stream into several parallel bit streams, each of which has a much lower bit rate, and by using these sub-streams to modulate several sub-carriers [1]–[2]. Multi-carrier modulation possesses several properties that make it an attractive approach for high-speed wireless communication networks. Among these properties is the ability to efficiently access and distribute multiplexed data streams, and a reduced susceptibility to impulsive as well as narrowband channel disturbances.

Each data symbol in single-carrier systems occupies the entire available bandwidth while an individual data symbol in multi-carrier system only occupies a fraction of the total bandwidth. Therefore, narrowband interference or strong frequency selective attenuation can cause single-carrier transmission to completely fail, but in MCM they only affect sub-carriers located at particular frequencies.

MCM can be implemented using several techniques. The first multi-carrier systems employed frequency division multiplexing (FDM). In FDM the composite multi-carrier signal is obtained by shifting the baseband parallel data streams upwards in frequency by modulating them on different sinusoidal carriers. The FDM signal must consist of sub-carriers that do not have overlapping spectra or otherwise crosstalk would occur between different sub-channels. In practical systems the guard bands are inserted between sub-carriers in order to accommodate for local oscillators imperfections and/or channel effects like Doppler spread. Figure 3.3a shows the spectrum of composite FDM signal with guard bands.

There is, however, an alternative approach to transmitting data over a multi-path channel. Instead of using carriers with non-overlapping bands, one could partition the spectrum into closely packed sub-bands that overlap. In the next sections we shall see how this is done to optimize utilization of the spectrum, a resource in premium.

3.2 Orthogonal frequency division multiplexing

Over the years there have been several approaches towards the realization of multi-carrier transmission, especially aiming at optimum utilization of spectral bandwidth. One of the spectrally efficient multi-carrier methods is orthogonal frequency division multiplexing (OFDM) [3]. Although the principle of OFDM has existed since the early 1960s, the first real-life systems appeared only in the 1990s. Today, OFDM is the most commonly used multi-carrier modulation technique and is widely adopted across the world. It is in fact the de-facto choice for high-speed data rate transmission in frequency-selective fading channels and wireless-local area networks (WLAN). One of the first systems to use OFDM was European Digital Audio Broadcasting (DAB) back in 1995 and in a short time other standards such as Digital Video Broadcasting (DVB), WiFi (IEEE 802.11a/g/j/n), WiMAX (IEEE 802.16), UWB Wireless PAN (IEEE 802.15.3a) and MBWA (IEEE 802.20) followed [1].

The high spectral efficiency of OFDM is due to its orthogonal sub-carriers, which allow their spectrums to overlap. Adjacent sub-carriers do not interfere with each other as long as they preserve their orthogonality. Furthermore, the guard bands like those used in FDM are no longer necessary; Figure 3.3b illustrates this with the spectrum of OFDM for 8 sub-carriers. The technique has other advantages, too – high immunity to multi-path delay spread that causes inter-symbol interference (ISI) in wireless channels, immunity to frequency selective fading channels, elegance in implementation through the fast Fourier transform (FFT) algorithms and ease of channel equalization.

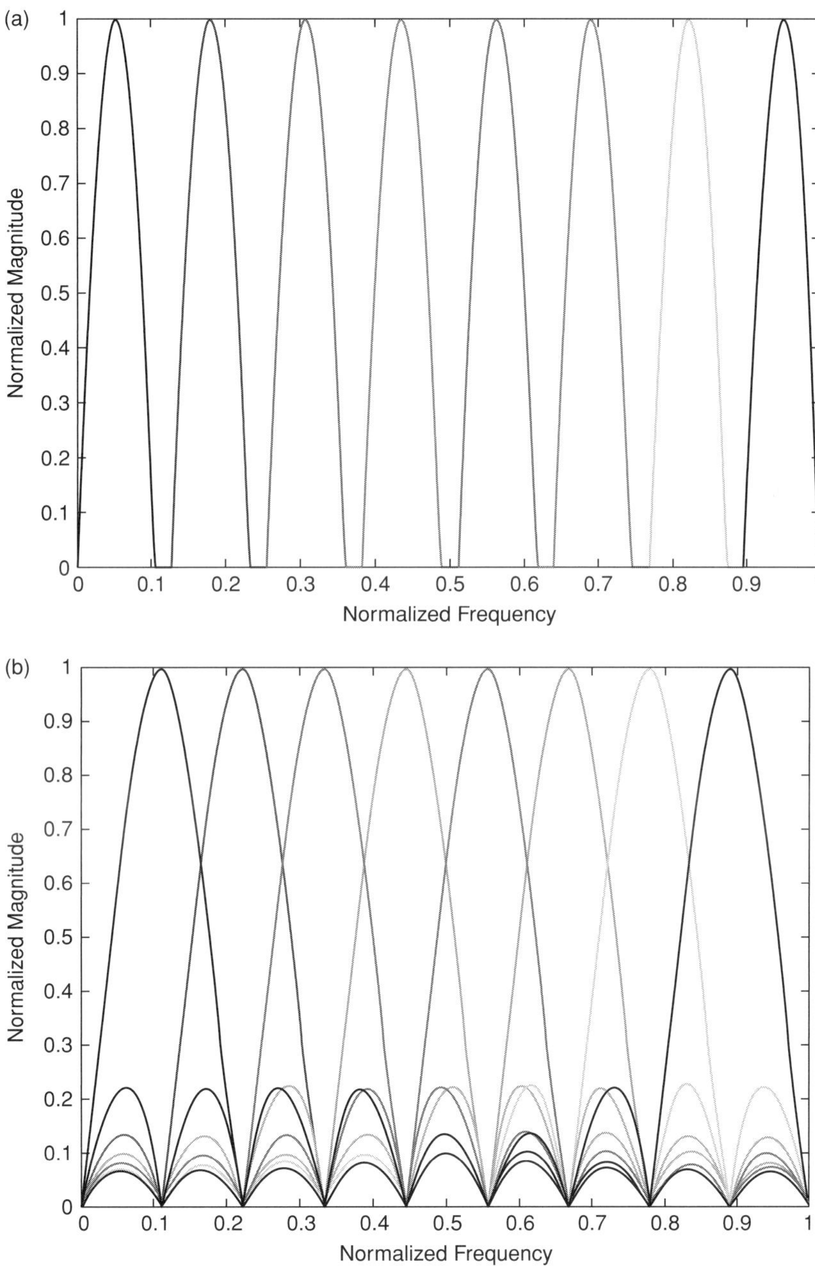

Figure 3.3 (a) FDM spectrum (8 sub-carriers with guard bands); (b) OFDM spectrum (8 sub-carriers).

OFDM transmission system can be efficiently implemented using the inverse fast Fourier transform (IFFT) at the transmitter side and fast Fourier transform (FFT) at the receiver side. The Fourier transformation allows us to describe a signal as a linear combination of sinusoids that form an orthogonal basis. These sinusoids in OFDM are referred to as sub-carriers and their number is determined by the length of the FFT vector. The orthogonality of sub-carriers over an OFDM symbol period T_{MC} is achieved by setting the inter-carrier spacing to $^1/T_{MC}$ Hz. Therefore, the frequency of the kth sub-carrier in so-called T-spaced OFDM is given by [3]:

$$f_k = \frac{k}{T_{MC}}, \quad k = 0, 1, \ldots, N-1 \tag{3.1}$$

The corresponding kth sub-carrier at frequency f_k can therefore be written as [3]:

$$\vartheta_k(t) = e^{j2\pi f_k t} \tag{3.2}$$

An OFDM symbol consists of N sub-carriers and after being modulated by the OFDM transmitter can be expressed as:

$$S[n] = \sum_{k=0}^{N-1} a_k e^{j2\pi \frac{kn}{N}}, \quad 0 \leq n \leq N-1 \tag{3.3}$$

In Eq. (3.3) a_k represents the mapped complex data symbols. If we assume an ideal channel and perfect synchronization between OFDM transmitter and receiver, the received sequence $R[n]$ is identical to the transmitted signal; i.e., $R[n] = S[n]$. Under such conditions the demodulated data after FFT for the kth sub-carrier can be expressed as:

$$\begin{aligned} \hat{a}_{k'} &= \frac{1}{N}\sum_{n=0}^{N-1} R[n] e^{-j2\pi \frac{k'n}{N}} \\ &= \frac{1}{N}\sum_{n=0}^{N-1}\sum_{k=0}^{N-1} a_k e^{j2\pi \frac{kn}{N}} e^{-j2\pi \frac{k'n}{N}} \\ &= \sum_{k=0}^{N-1} a_k \left(\frac{1}{N}\sum_{n=0}^{N-1} e^{j2\pi \frac{n(k-k')}{N}}\right) \\ &= \sum_{k=0}^{N-1} a_k \delta[k-k'] = a_k \end{aligned} \tag{3.4}$$

Unfortunately, the above scenario describes an idealistic situation that does not occur often in reality and therefore the channel effects and oscillators' imperfections should be taken into consideration during system design. Due to delay spread of the channel, OFDM symbols could overlap one another and perfect reconstruction as described in Eq. (3.4) may not be possible. In order to decrease the amount of ISI in dispersive channels, guard intervals are inserted between OFDM symbols. Usually in OFDM the cyclic prefix is used as it makes the OFDM signal appear periodic and therefore avoid the discrete time property of the convolution.

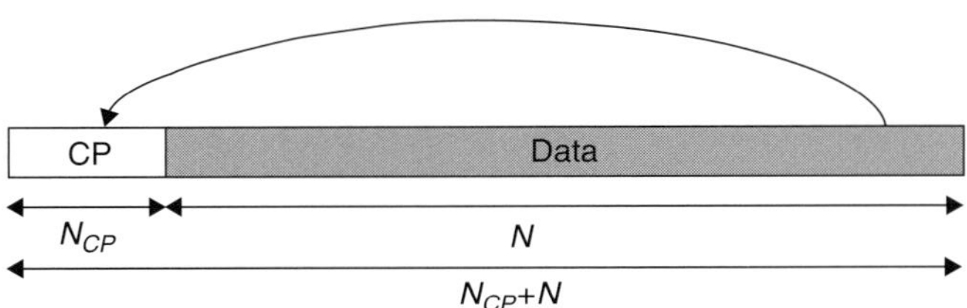

Figure 3.4 OFDM symbol with cyclic prefix.

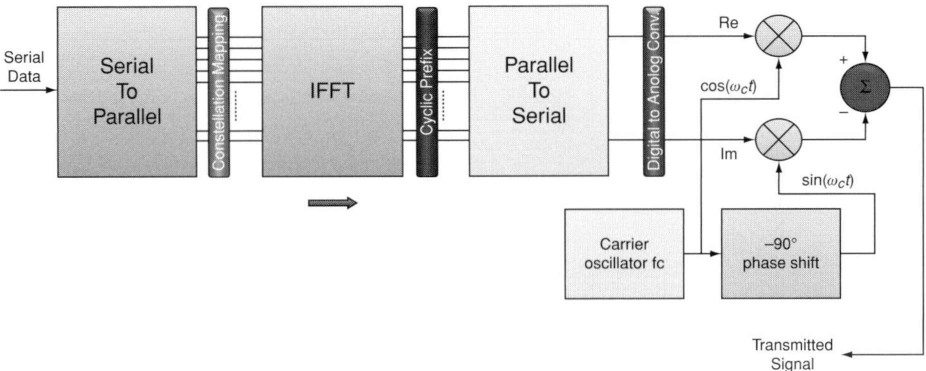

Figure 3.5 OFDM transmitter.

The cyclic prefix is a copy of last N_{CP} samples of OFDM symbols that is appended to the front of each symbol. The effect of the dispersive channels can be efficiently mitigated if the length of a cyclic prefix is set longer than the span of the channel. Figure 3.4 depicts an OFDM symbol with cyclic prefix. Because the cyclic prefix does not contain any useable data, it decreases the spectral efficiency and therefore it is kept as short as possible. At the receiver side the cyclic prefix is no longer needed and hence discarded before the demodulation process.

The OFDM transmitter and receiver block diagrams are illustrated in Figures 3.5 and 3.6, respectively.

3.3 Filter bank multi-carrier methods

OFDM uses carriers with the characteristics of a *sinc*-function. This has some desirable signal orthogonality properties, namely zero inter-symbol interference (ISI) and zero inter-carrier interference (ICI) in spite of the carriers overlapping. However, the *sinc* characteristic causes the sub-carriers to have large side-lobes that spill over into neighbouring bands resulting in significant interference. Furthermore, under non-ideal channel conditions, the spectral overlap between the sub-channels necessitates the use of cyclic prefix (CP) and frequency offset correction algorithms. Although the CP is an easy solution to mitigate the impairments induced by the channel, it leads to a loss in the data throughput and bandwidth efficiency. There exist in the literature several alternative

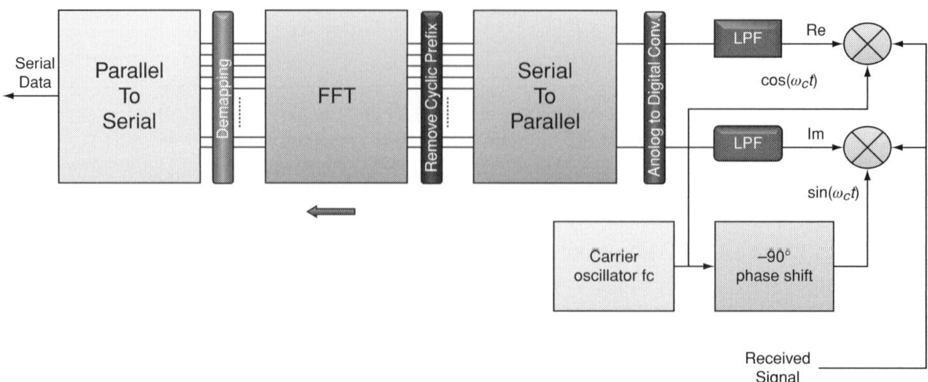

Figure 3.6 OFDM receiver.

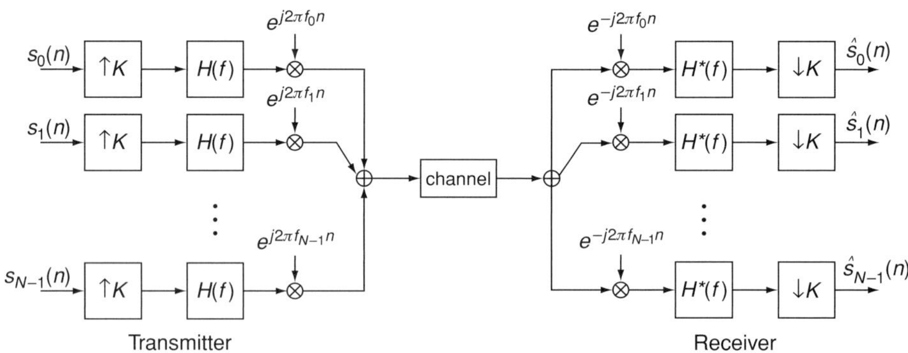

Figure 3.7 FMT transmitter and receiver, after [5].

multi-carrier techniques to OFDM [4]–[5]. We shall discuss a few important ones in the next sub-sections.

3.3.1 Filtered multi-tone (FMT)

In [6]–[7] a filter-bank modulation technique called filtered multi-tone (FMT) is presented. FMT is similar to frequency division multiplexing (FDM) in the sense that the sub-carriers do not overlap and guard bands are used between carriers to prevent interference. FMT is implemented using filter banks with a single prototype filter $H(f)$ and its dual $H^*(f)$. The prototype filter is usually a root-Nyquist filter [6]. The modulation scheme is usually quadrature amplitude multiplexing (QAM). Figure 3.7 shows the implementation of the FMT transmitter and the FMT receiver modulator. In the figure, N denotes the maximum number of sub-carriers and K represents the sampling factor. Usually a choice of $K > N$ is made for addition of guard bands between the sub-carrier bands. Equalizers are needed after down-sampling at the receiver.

In FMT, orthogonality between sub-channels is ensured by using non-overlapping spectral characteristics as compared with the overlapping *sinc*-function-type spectra

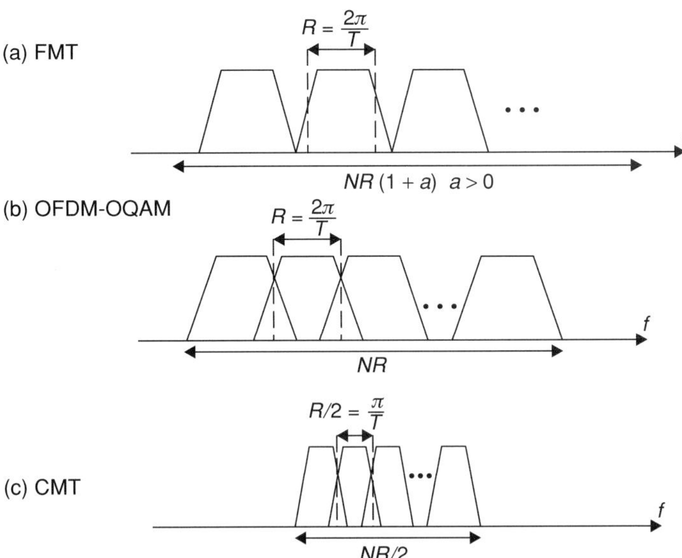

Figure 3.8 Sub-carrier signal spectra of (a) FMT, (b) OFDM-OQAM and (c) CMT.

employed in OFDM. Since the linear transmission medium does not destroy orthogonality achieved in this manner, cyclic prefixing is not needed. Clearly, the required amount of spectral containment must be achieved with acceptable filtering complexity. In a critically sampled ($N = K$) filter bank, the frequency separation of the pass bands will be $1/T$ with a total of M bands. In this way, each of the transmitter pass-band filters will be frequency-shifted versions of the low-pass filter as shown in Figure 3.8a. An obvious disadvantage of FMT is the inefficient use of bandwidth as the sub-carriers do not overlap.

3.3.2 Cosine modulated multi-tone (CMT)

In [8] Farhang and Boroujeny introduces the cosine modulated multi-tone (CMT) as a capable multi-carrier modulation technique. Figure 3.9 shows the blocks of the CMT transmitter and receiver. The CMT uses pulse amplitude modulated (PAM) symbols with vestigial sideband modulation and the sub-carrier bands are maximally overlapped/minimally spaced. In order to allow maximum bandwidth efficiency, vestigial sideband modulation (illustrated in Figure 3.10) is adopted. Root-Nyquist filters are selected for $H(f)$ and $H^*(f)$ to ensure separation of data symbols at the receiver. Equalizers are used after decimators at the receiver [8].

Both FMT and cosine modulated multi-tone (CMT) are filter-bank-based modulation techniques [8]. The main difference between the two methods lies in the way the spectral band is used, as shown in Figures 3.8a and 3.8c. In FMT, the sub-carrier bands are non-overlapping, and thus separation of different sub-carrier signals can be achieved by conventional filtering. On the other hand, in CMT, the sub-carrier bands are allowed to overlap and separation is done through judicious design of the synthesis and analysis

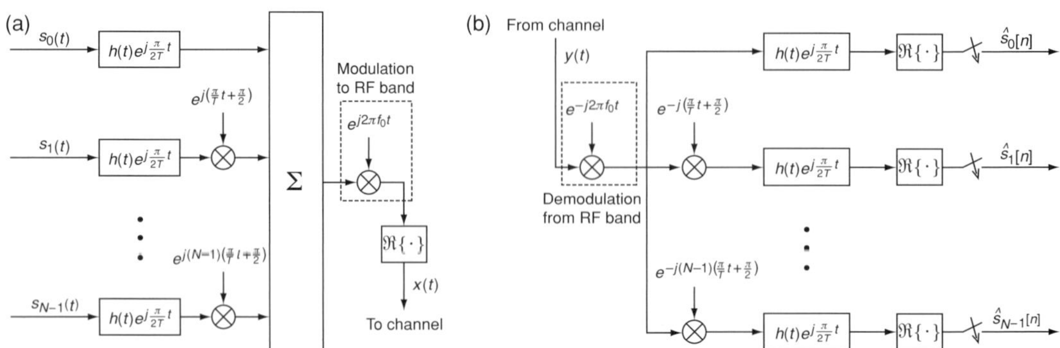

Figure 3.9 CMT. (a) Transmitter and (b) receiver.

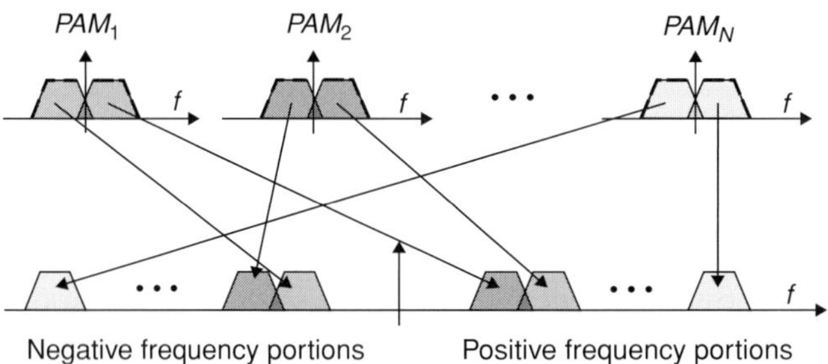

Figure 3.10 Illustration of vestigial side-band modulation of the CMT carriers, after [5].

filters. It is clear from Figure 3.8c that CMT offers higher bandwidth efficiency than FMT, since more sub-carrier bands can be accommodated per unit bandwidth.

In both FMT and CMT, the sub-carriers can be considered to be of narrow bandwidth experiencing a flat fading channel. Hence, equalization after sub-carrier separation can be established through a single tap equalizer, whose tap weight is the inverse of the channel gain. Training symbols are usually used to initialize the equalizer taps. In CMT, the very special structure of the underlying signals allows for blind equalization without training. The procedures are described in [8].

3.3.3 OFDM-offset QAM/staggered multi-tone (SMT)

Another technique suggested is the staggered multi-tone (SMT) modulation. The method is also known as offset QAM and is implemented using poly-phase filter banks [9]–[10]. Unlike FMT, the method SMT allows overlap of carriers to maximize spectrum utilization. The modulation scheme used is offset QAM, where the quadrature and in-phase components are separated by a time offset of half the symbol interval, hence the name staggered multi-tone.

In OFDM-OQAM, the sub-carrier bands overlap and are spaced at the symbol rate. Successful signal separation is nevertheless possible thanks to the orthogonality condition between sub-carriers that guarantees that the transmitted symbols arrive at the

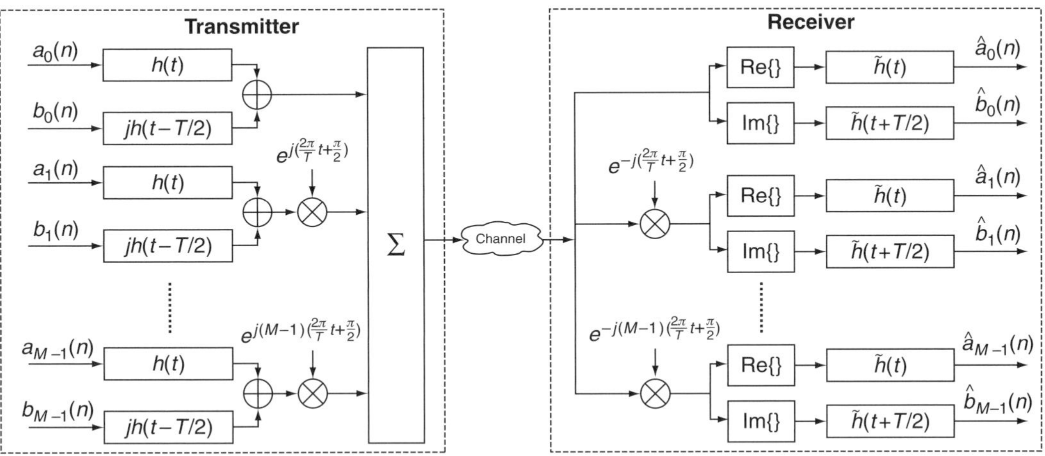

Figure 3.11 OFDM-OQAM transmitter and receiver, after [5].

receiver free of inter-symbol (ISI) and inter-carrier interference (ICI). Carrier orthogonality is achieved through time staggering the in-phase and quadrature components of the sub-carrier symbols and designing proper transmit and receive filters. In OFDM-OQAM, each sub-carrier band is double sideband modulated and carries a sequence of QAM (complex valued) symbols. In contrast, the sub-carrier modulation in CMT is vestigial sideband where the sub-carriers modulate real-valued PAM symbols. Assuming identical symbol duration and number of sub-carriers, the CMT signal occupies half the bandwidth of OFDM-OQAM; thus, only providing half of its data rate. FMT, on the other hand, introduces guard bands between adjacent sub-carriers that are complex modulated. The width of the guardbands depends on the specific system implementation. Therefore, for an identical number of carriers and identical symbol timing, FMT requires more bandwidth than OFDM-OQAM and CMT [9].

The OFDM-OQAM method is similar to CMT for the case when the sub-carrier bands are maximally overlapping (i.e., are minimally spaced), see Figure 3.8b. Both OQAM and CMT achieve maximum bandwidth efficiency. Transmit symbols of OFDM-OQAM are offset QAM: in-phase and quadrature components have a time offset of half the symbol interval. If the overlaps are limited to adjacent bands and $H(f)$ and $H^*(f)$ are a pair of root-Nyquist filters, the separation of data symbols at the receiver output is guaranteed. Equalizers are needed after decimators at the receiver. Figure 3.11 shows the blocks of the OFDM-OQAM transmitter and receiver.

3.4 Wavelet and wavelet-packet-based multi-carrier modulators

3.4.1 Wavelet packet modulator (WPM)

Recently, the theory of wavelets [11] and wavelet packets [12] has been applied for the design of multi-carrier modulators. The pioneering studies on these subject were carried out by Lindsay [13], who laid out the theoretical foundations to establish the link between wavelet packets and digital communication. He also showed that the entire WPM transceiver structure can be realized with a pair of conjugate quadrature mirror filters that satisfy a set of constraints. His idea has since then been taken forward by many researchers. The decoding of WPM data with maximum likelihood estimators has been addressed by Suzuki et al. [14]. The study of an equalization scheme suited for WPM has been conducted by Gracias and Reddy [15]. In [16]–[17] an investigation on the performance of WPM systems in the presence of time offset is performed. In [18] its PAPR performances are analyzed. The advantages of the wavelet transform in terms of the flexibility they offer to customize and shape the characteristics of the waveforms have been demonstrated in [19]–[22]. Three use-cases where the waveforms are designed and applied to optimize the WPM system performance according to specific system demands are illustrated in [19]–[21]. In [22], the work of [19]–[20] is extended to establish a unifying mathematical framework where the waveforms are designed according to a pre-defined criteria.

WPM is implemented with orthogonal wavelet packet (WP) bases derived from a multi-resolution analysis (MRA). Fundamentally, OFDM and WPM have many similarities as both use orthogonal sub-carriers (which overlap one another) to achieve high spectral efficiency. The adjacent sub-carriers do not interfere with each other as long as the orthogonality between sub-carriers is preserved. The difference between OFDM and WPM is the time–frequency characteristics of the sub-carriers and in the manner they are generated. OFDM uses Fourier bases that are static sines/cosines, while WPM uses wavelets that offer more flexibility. By utilization of different wavelets in WPM, we can get different sub-carriers which lead to different transmission system characteristics. Therefore, by proper selection of wavelets, it is possible in WPM to optimize figures of metrics like bandwidth utilization, sensitivity to synchronization errors, peak-to-average power ratio (PAPR), etc.

The starting point to derive the orthogonal bases is to consider a pair of quadrature mirror filters (QMF) consisting of a half-band low-pass filter $h[n]$ and high-pass filter $g[n]$ of length L each. These filters share a tight relationship given by [23]–[24]:

$$g[L-1-n] = (-1)^n h[n] \tag{3.5}$$

Furthermore, they have adjoints or duals that are their complex conjugate time-reversed variants [24]:

$$h'[n] = h^*[-n] \text{ and } g'[n] = g^*[-n] \tag{3.6}$$

The filter-pair $\{h'[n], g'[n]\}$ are called the synthesis filters and are used to generate the WP carriers for modulation of the data at the transmitter. On the other hand, the combination $\{h[n], g[n]\}$ is called the analysis filter and is used to derive the duals for demodulation of data at the receiver. Denoting the magnitude responses of these four filters in the frequency domain as $H(\omega)$, $G(\omega)$, $H'(\omega)$ and $G'(\omega)$, the filters are said to satisfy the perfect reconstruction conditions if [24]:

$$\begin{aligned} &H^*(\omega+\pi)H'(\omega) + G^*(\omega+\pi)G'(\omega) = 0 \\ &H^*(\omega)H'(\omega) + G^*(\omega)G'(\omega) = 2 \end{aligned} \tag{3.7}$$

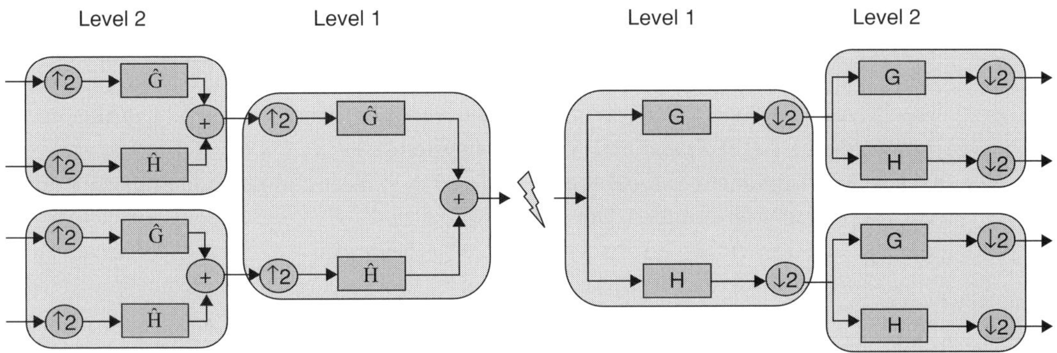

Figure 3.12 Wavelet-packed-based transmultiplexer.

Such filters can be used for various applications from compression of image/speech signals to radio system design. From these QMF filters, the wavelet packet bases $\{\xi_l^p\}$ can be derived recursively through an MRA as [24]:

$$\begin{aligned}\xi_{l+1}^{2p}[n] &= \sqrt{2}\sum_k h[k]\xi_l^p[2n-k]\\ \xi_{l+1}^{2p+1}[n] &= \sqrt{2}\sum_k g[k]\xi_l^p[2n-k]\end{aligned} \tag{3.8}$$

In Eq. (3.8) ξ denotes the wavelet packets duals and p stands for the sub-carrier index at any given tree depth l. The decomposition levels in the wavelet packet tree determine the number of WPM sub-carriers:

$$N = 2^l \tag{3.9}$$

In Eq. (3.9) N stands for the number of sub-carriers and l represents the number of levels in the filter bank. These bases satisfy two orthogonal properties that are crucial for their application to MCM. First, they are orthogonal to themselves for all non-zero integer shifts, i.e. [25]:

$$\langle \xi_l^p[n-j], \xi_l^p[n-k]\rangle = \delta[j-k], \quad \forall j,k \in \mathbb{Z} \tag{3.10}$$

Secondly, pairs of the WP bases derived out of the same parent are orthogonal to one another for all j and k [25]:

$$\langle \xi_l^{2p}[n-j], \xi_l^{2p+1}[n-k]\rangle = 0, \quad \forall j,k \in \mathbb{Z} \tag{3.11}$$

Equation (3.8) can be physically realized with a filter bank tree structure obtained by cascading the fundamental $\{h[n], g[n]\}$ filter pair, followed by down-sampling by 2, iteratively as shown under the discrete wavelet packet transform (DWPT) block in Figure 3.12 [25].

The figure shows a level-2 decomposition scheme that yields up to 4 orthogonal WP bases. The WP duals $\{\xi_l'^p\}$ for the transmitter can be obtained by a similar procedure, albeit with the synthesis filter pairs $\{h'[n], g'[n]\}$. The processes are referred to as inverse-DWPT (IDWPT) and DWPT at the transmitter and receiver, respectively, analogous to the inverse-FFT (IFFT) and FFT, in OFDM systems.

The WPM-modulated signal $S[n]$ is obtained as a linear combination of the WP duals $\{\xi_l'^p\}$ weighted with complex data symbols $a_{u,k}$:

$$S[n] = \sum_u \sum_{k=0}^{N-1} a_{u,k} \xi_l'^k [n - uN] \tag{3.12}$$

In Eq. (3.12) k denotes the sub-carrier index and u denotes the WPM symbol index. The constellation symbol modulating kth sub-carrier in the uth WPM symbol is represented by $a_{u,k}$.

At the receiver, the data are demodulated with the dual bases. If we assume that the WPM transmitter and receiver are perfectly synchronized and that the channel is ideal, the detected data at the receiver can be given by:

$$\begin{aligned} \hat{a}_{u',k'} &= \sum_n R[n] \xi_l^{k'} [u'N - n] \\ &= \sum_n \sum_u \sum_{k=0}^{N-1} a_{u,k} \xi_l^k [n - uN] \xi_l^{k'} [u'N - n] \\ &= \sum_u \sum_{k=0}^{N-1} a_{u,k} \left(\sum_n \xi_l^k [n - uN] \xi_l^{k'} [u'N - n] \right) \\ &= \sum_u \sum_{k=0}^{N-1} a_{u,k} \delta[u - u'][k - k'] = a_{u,k} \end{aligned} \tag{3.13}$$

An important property unique to wavelet transform is that the wavelet bases are much longer than the duration of a symbol and can overlap in the time domain without losing their orthogonality. The longer waveforms in WPM allow for better frequency localization of sub-carriers, especially in relation to OFDM where the rectangular DFT windows result in large side-lobes. In Figure 3.13 the spectrum of a WPM system with 8 sub-carriers is depicted.

An undesirable consequence of time overlap is the inability to use a guard band in WPM systems. Although the addition of a guard interval in OFDM severely decreases spectral efficiency, it is an effective and low-complexity method to cope with dispersive channels and time offsets.

The WPM transmitter and receiver block diagrams are illustrated in Figures 3.14a and 3.14b, respectively.

3.4.2 Variants of wavelet packet modulator

The wavelet packet modulator can be considered as a generalized form of other multi-carrier modulators based on wavelets. In [11] Negash and Nikookar suggest replacing

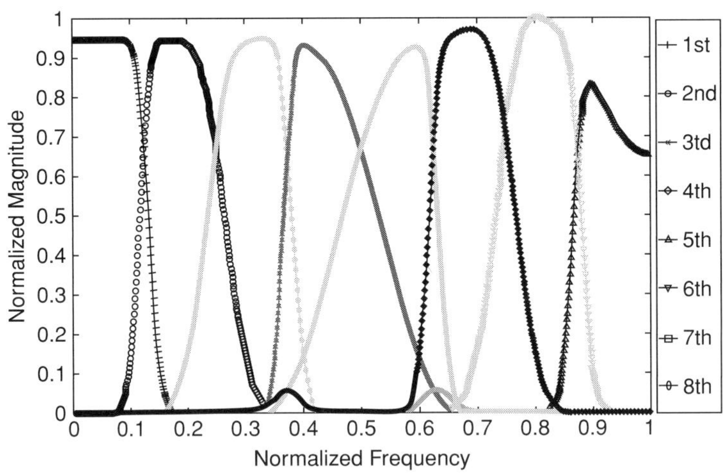

Figure 3.13 Spectrum of 8 WPM orthogonal subcarriers (for Daubechies filter of length 20).

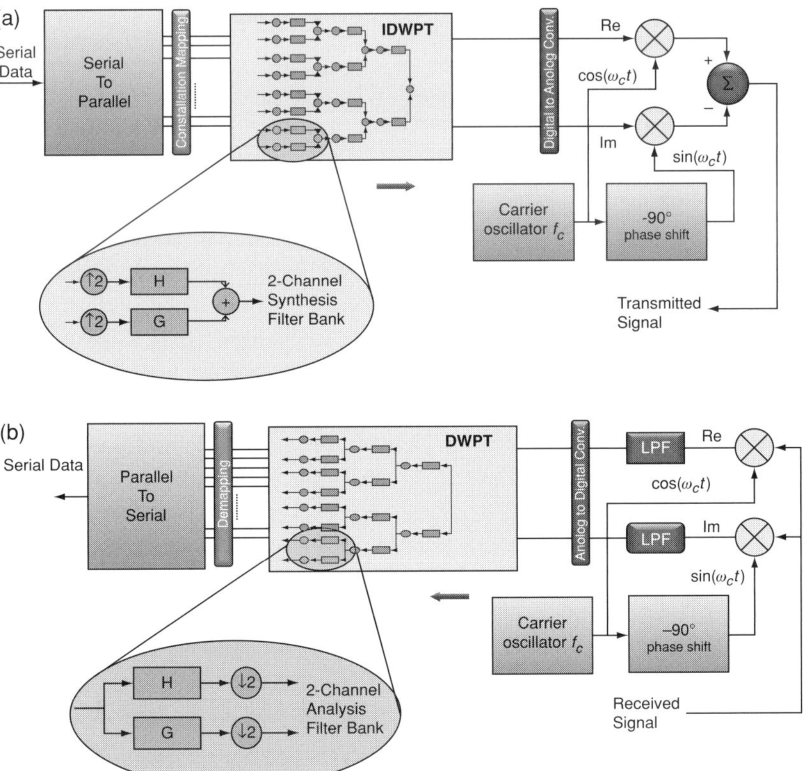

Figure 3.14 (a) WPM transmitter; (b) WPM receiver.

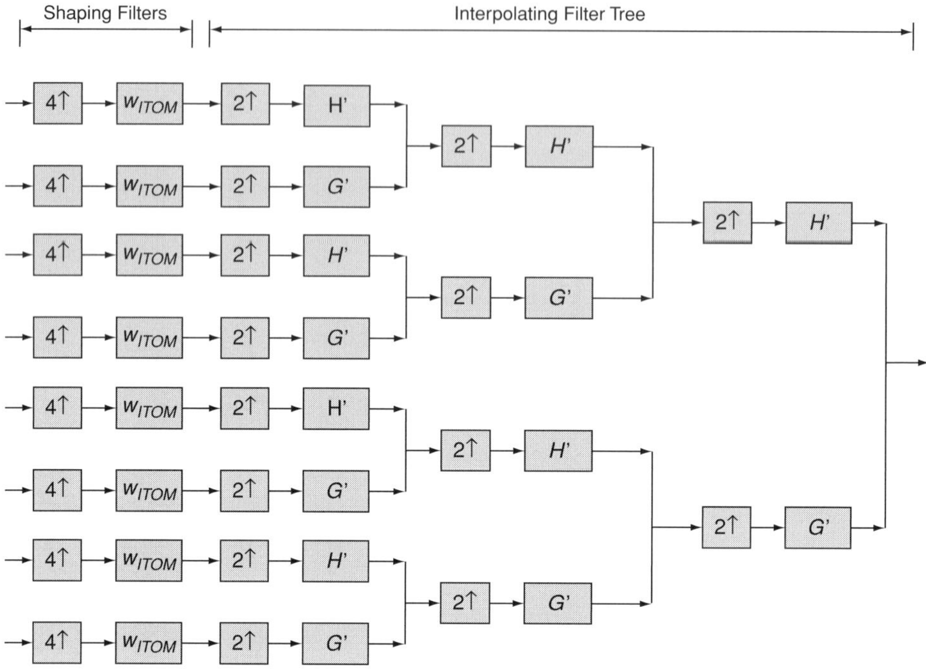

Figure 3.15 Modulator of interpolated tree structure (ITOM). In the figure w_{ITOM} is the ITOM shaping filters.

the conventional Fourier-based complex exponential carriers of a multi-carrier system with orthonormal wavelets. The wavelets are derived from a multi-stage tree-structured Haar and Daubechies orthonormal QMF bank. An improved performance with respect to reduction of the power of ISI and ICI is reported. This work is extended in [28] by realizing a high-speed digital communication system over a low-voltage power-line. With empirical investigations on a model obtained from measurements of a practical low-voltage powerline communication channel, the authors demonstrate the effectiveness of wavelets for use in OFDM systems, especially with regard to ISI and ICI mitigation. Another real-time application of the system is reported in [29] where wavelet-based OFDM for V-BLAST (vertical Bell laboratories layered space time) [30] is discussed. According to [29] the bit error rate (BER) performance of the wavelet-based V-BLAST system is superior to their Fourier-based counterparts. In the conventional systems, the ISI and ICI are reduced by adding a guard interval (GI) using a cyclic prefix (CP) to the head of the OFDM symbol. Adding CP can largely reduce the spectrum efficiency. Wavelet-based OFDM schemes do not require CP, thereby enhancing the spectrum efficiency. Moreover, as pilot tones are not necessary for the wavelet-based OFDM system, they perform better in comparison to existing OFDM systems like 802.11a or HiperLAN, where 4 out of 52 sub-bands are used for pilots. An advanced OFDM modulation scheme called the isotropic orthogonal transform algorithm (IOTA)

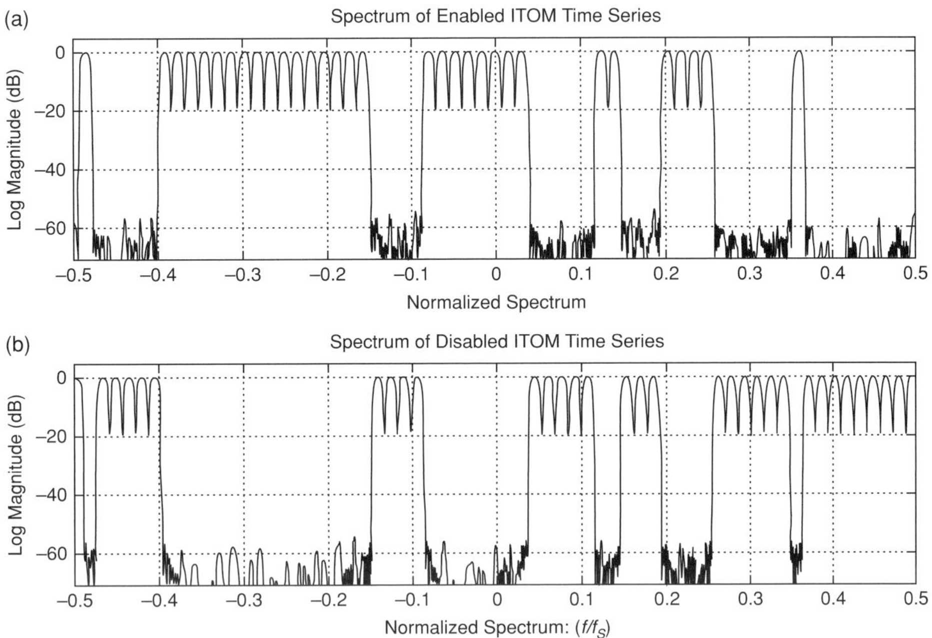

Fig. 3.16 Illustration of ITOM operation: (a) Spectra of enabled spectral bands of 64-point ITOM; (b) spectra of disabled spectral bands of 64-point ITOM, after [26].

for future broadband physical layers is proposed in [31]. This system uses isotropic Gaussian functions to generate the carrier waves and gives good spectral efficiency by eliminating the use of a cyclic prefix. In [32] the promise shown by a Haar WOFDM system with Hadamard spreading codes in reducing its peak-to-average power ratio (PAPR) is reported.

3.4.3 Interpolated tree orthogonal multiplexing (ITOM)

In the WPM technique the filter banks perform the dual role of shaping the spectrum as well as interpolating in time series. A slight enhancement to this approach would be to separate the two processes and gain greater control over the characteristics of the carriers. This method is called the interpolated tree orthogonal multiplexing (ITOM) and was introduced by Harris and Kjeldsen [26]. The procedure is depicted in Figure 3.15. From the figure we can notice that up-sampled shaping filters precede the input ports of the wavelet packet tree structure. Notching over the desired spectral interval is achieved by vacating one or more of the input branches. Figure 3.16 illustrates an example of the ITOM mechanism. We may note from Figures 3.16(a) and (b) how well the enabled and disabled carries fit into the spectral gaps of one another, illustrating the superiority of the ITOM procedure towards spectrum shaping.

3.5 Summary

In this chapter we discussed various multi-carrier techniques available for efficient modulation of data. OFDM was presented as the most popular of MCM implementations. Filter bank alternatives to OFDM, like FMT, CMT and SMT, were addressed. The operation of the WPM transceiver, as a wavelet-based implementation of orthogonal multi-carrier system, was presented.

The WPM is a relatively young multi-carrier transmission technique and very little is known about its operation. In the next two chapters we shall evaluate the operation of the WPM system over various performance metrics like:

- sensitivity to loss of synchronization (time/frequency/phase);
- peak-to-average power ratio (PAPR) performance.

References

[1] H. Nikookar, *Advanced Topics in Digital Wireless Communications*, published by IRCTR-Delft University of Technology, 2004.

[2] J.G. Proakis, *Digital Communications*, 2nd edn, McGraw-Hill Book Company, 1989.

[3] R. van Nee and R. Prasad, *OFDM for Wireless Multimedia Communications*, Artech House Publishers, 2000.

[4] P. Amini, R. Kempter and B. Farhang-Boroujeny, "A Comparison of Alternative Filterbank Multicarrier Methods for Cognitive Radio Systems," *Proceedings of the SDR 06 Technical Conference and Product Exposition*, 2006.

[5] B. Farhang-Boroujeny, Signal Processing Techniques for Spectrum Sensing and Communications in Cognitive Radios, Online Tutorial Source, www.crowncom.org/2007/t2.pdf.

[6] G. Cherubini, E. Eleftheriou, and S. Olcer (1999), "Filtered Multitone Modulation for VDSL," in *Proc. IEEE Globecom 99*, vol. 2, pp. 1139–44, 1999.

[7] P. Amini, R. Kempter, R.R. Chen, L. Lin and B. Farhang-Boroujeny, "Filter Bank Multitone: A Physical Layer Candidate For Cognitive Radios," *SDR Forum Technical Conference*, Hyatt Regency, Orange County, California, 14–17 November 2005.

[8] B. Farhang-Boroujeny, "Multicarrier Modulation with Blind Detection Capability Using Cosine Modulated Filter Banks," *IEEE Trans. Commun.*, vol. 51, no. 12, Dec. 2003, pp. 2057–70.

[9] B.R. Saltzberg, "Performance of an Efficient Parallel Data Transmission System," *IEEE Trans. Comm. Tech.*, vol. 15, no. 6, pp. 805–11, Dec. 1967.

[10] B. Hirosaki, "An Orthogonally Multiplexed QAM System Using the Discrete Fourier Transform," *IEE Transactions on Communications*, vol. Com 29, No. 7, July 1981.

[11] B.G. Negash and H. Nikookar, "Wavelet-based Multicarrier Transmission over Multipath Wireless Channels," *IEE Electronics Letters*, vol. 36, No. 21, pp. 1787–8, October 2000.

[12] A. Jamin and P. Mahonen, "Wavelet Packet Modulation for Wireless Communications," *Wireless Communications & Mobile Computing Journal*, vol. 5, Issue 2, John Wiley, March 2005.

[13] A. Lindsay, "Wavelet Packet Modulation for Orthogonally Transmultiplexed Communications," *IEEE Transactions on Signal Processing*, vol. 45, pp. 1336–9, May 1997.

[14] N. Suzuki, M. Fujimoto, T. Shibata, N. Itoh, and K. Nishikawa, "Maximum Likelihood Decoding for Wavelet Packet Modulation," *IEEE Vehicular Technology Conference (VTC)*, pp. 2895–8, 1999.

[15] S. Gracias and V.U. Reddy, "An Equalization Algorithm for Wavelet Packet based Modulation Schemes," *IEEE Transactions on Signal Processing*, vol. 46, pp. 3082–7, November 1998.

[16] H. Nikookar, "Comparison of Sensitivity of OFDM and Wavelet Packet Modulation to Time Synchronization Error," *PIMRC*, Cannes, Sept. 2008.

[17] D. Karamehmedović, M.K. Lakshmanan, and H. Nikookar, "Performance Degradation Study of Wavelet Packet based Multicarrier Modulation under Time Synchronization Errors," *WPMC '08* Lapland, Sept. 2008.

[18] M.K. Lakshmanan, B. Torun and H. Nikookar, "Peak-to-average Power Ratio Studies for Wavelet Packet Modulation," *European Wireless Technology Conference*, Rome, Italy, September 2009.

[19] D. Karamehmedovic, M.K. Lakshmanan, and H. Nikookar, "Optimal Wavelet Design for Multicarrier Modulation with Time Synchronization Error," *52nd Global Communications Conference (GLOBECOM)*, November 2009, Hawaii, USA.

[20] M.K. Lakshmanan and H. Nikookar, "Construction of Optimal Bases for Wavelet Packet Modulation under Phase Offset Errors," *3rd European Wireless Technology Conference*, Paris, France, September 2010.

[21] D.D. Ariananda, M.K. Lakshmanan, and H. Nikookar, "Design of Best Wavelet Packet Bases for Spectrum Estimation," *20th Memorial Symposium of Personal, Indoor and Mobile Radio Communications (PIMRC)*, Tokyo, Japan, September 2009.

[22] M.K. Lakshmanan, H. Nikookar, D. Karamehmedovic, and D.D. Ariananda, "A Unified Framework to Design Orthonormal Wavelets Packet Bases for Multi-Carrier Modulation," *Proc. IEEE International Conference on Communications (ICC) 2010*, Cape Town, South Africa, May 2010.

[23] M. Vetterli and I. Kovacevic, *Wavelets and Subband Coding*, Englewood Cliffs, New Jersey: Prentice-Hall PTR, 1995.

[24] P.P. Vaidyanathan, *Multirate Systems and Filter Banks*, Englewood Cliffs, NJ, Prentice-Hall, Inc., 1993.

[25] C.S. Burns, R.A. Gopinath, and H. Guo, *Introduction to Wavelets and Wavelet Transform, a Primer*, Prentice Hall, 1998.

[26] F. Harris and E. Kjeldsen, "A Novel Interpolated Tree Orthogonal Multiplexing (ITOM) Scheme with Compact Time-Frequency Localization: an Introduction and Comparison to Wavelet Filter Banks and Polyphase Filter Banks," *Proceedings of Software Defined Radio Technical Conference and Product Exposition*, Orlando, USA, November 2006.

[27] N. Erdol, F. Bao, and Z. Chen, "Wavelet Modulation: A Prototype for Digital Communication Systems," in *IEEE Southcon Conference*, pp. 168–71, 1995.

[28] Y. Zhang and S. Cheng, "A Novel Multicarrier Signal Transmission System over Multipath Channel of Low-voltage Power Line," *IEEE Transactions on Power Delivery*, Vol. 19, no. 4, pp. 1668–72, 2004.

[29] Z. Rafique, N. Gohar and M.J. Mughal, "Performance Comparison of OFDM and WOFDM based V-BLAST Wireless Systems," *IEICE Transactions on Communications*, Vol. E88-B, no. 5, pp. 2207–9, 2005.

[30] P.W. Wolniansky, G.J. Foschini, G.D. Golden, and R.A. Valenzuela, "V- BLAST: An Architecture for Realizing Very High Data Rates over the Rich-scattering Wireless Channel," *Proc. ISSSE-98*, Pisa, Italy, September 1998.

[31] D. Lacroix, J.P. Javaudin, and N. Goudard, "IOTA, An Advanced OFDM Modulation for Future Broadband Physical Layers," *7th Wireless World Research Forum (WWRF) Meeting*, Eindhoven, Netherlands, December 2002.

[32] K. Anwar, A.U. Priantoro, K. Ando, M. Saito, T. Hara, M. Okada and H. Yamamoto, "PAPR Reduction of OFDM Signals using Iterative Processing and Carrier Interferometry Codes," *IEEE Int. Symposium on Intelligent Signal Processing and Communications System (ISPACS 2004)*, pp. 48–51, Seoul, Korea, November 2004.

4 Synchronization issues of wavelet radio

4.1 Introduction

Multi-carrier modulation (MCM) techniques have gained precedence over other methods in the last decade. This is largely stimulated by the demands of higher bandwidth, increased use of wireless systems in time-dispersive environments (e.g., home, office, etc.) and lower costs of digital signal processing components; a requirement sufficiently catered to by MCM systems. It is a spectral efficient modulation scheme that divides the incoming high rate data among multiple carriers modulated at lower rates. By simultaneously transmitting N data symbols into N carriers the symbol rate is reduced to the one Nth of the original symbol rate, and the symbol duration is increased N-fold. This leads to a transmission system that is robust against channel dispersions/fading, impulse noise and multi-path interference[1].

The rapid increase in wireless applications and the ensuing lack of free spectrum have prompted engineers to pursue bandwidth-efficient multi-carrier techniques. In order to achieve high bandwidth efficiency, the sub-carriers have to be closely spaced to each other. In this class of multi-carrier systems belong OFDM and WPM [1]–[3], discussed in Sections 3.2 and 3.4, respectively. OFDM and WPM have orthogonal sub-carriers that overlap with one another. The orthogonality property of the transmission bases ensures that the information containing sub-carriers does not interfere. Before the MCM symbol can be demodulated, the receiver has to be synchronized properly with the transmitted frame timing, carrier frequency and phase. However, impairments such as frequency offset and/or phase noise, induced by radio front ends or channel conditions, can cause the sub-carriers to lose their mutual orthogonality and impede the transmission of one another. The rise of interference level due to loss of orthogonality is far more pronounced in multi-carrier transmission than in single-carrier systems. This disadvantage of multi-carrier systems places high demands on the quality of the analogue radio components, especially on the choice of oscillators. For OFDM transmission the effects of frequency offset and phase noise are well documented in the literature [1]–[10] and a number of synchronization techniques are reported to estimate and reduce the frequency offset and phase noise effects [11]–[17]. Similar material for WPM performance does not exist.

Besides frequency and phase misalignment, multi-carrier systems can also suffer from loss of time synchronization. Time synchronization errors occur when the start of the

[1] Increased symbol duration leads to better performance in dispersive channels. Use of several narrow sub-bands ensures that narrowband interference or strong frequency band attenuation only affects particular sub-carriers and not the whole system.

multi-carrier symbol is incorrectly detected; this may happen when selecting parts of the adjacent symbol while discarding samples at the beginning or end of the useful symbol. Due to loss of time synchronization, disturbances like inter-symbol interference (ISI) and inter-carrier interference (ICI) occur. The use of guard intervals, like cyclic prefix in OFDM, can significantly improve the system performance in case of timing errors. However, the use of guard intervals is not a feasible solution for WPM systems because the wavelet packet transform is a lapped transform. As with the study of the impact of frequency offset and phase noise effects on the performance of the multi-carrier systems, so with the effects of timing errors the literature is virtually non-existent for WPM relative to OFDM. The sensitivity of the OFDM to the time synchronization error is reported in [18]–[20] and there are various techniques for OFDM symbol synchronization available in the literature [21]–[25].

In this chapter we address the impact of different synchronization errors on the WPM transmission and compare their performances with that of OFDM. The operation of WPM transceivers, employing different wavelets, is numerically evaluated under different conditions. Analytical expressions for the demodulation of the transmitted WPM data are also derived. Each of the frequency, phase and time errors is treated individually and separately in different sub-sections. The idea to study the three disturbances separately is in part to gain a better understanding of the individual phenomenon but also to aid ease of analysis. Moreover, the three errors – time, frequency and phase – are caused by disparate processes and the approach to study them separately is a reasonable approximation. First, we present the impact of carrier frequency offsets on WPM/OFDM communication in Section 4.2. This is followed by a discussion on the influence of phase noise in Section 4.3. Lastly, the transmission of WPM/OFDM under a loss of time synchronization error is analyzed in Section 4.4. The chapter rounds off with a summary in Section 4.5.

4.2 Frequency offset in multi-carrier modulation

The orthogonality between sub-carriers is maintained as long as the transmitter and receiver have the same reference frequency. Any offset in the frequency results in loss of orthogonality and the carriers interfere with one another's transmission. This is because, during demodulation, sampling may not occur at the peaks of the sub-carriers but rather at offset points. Besides the interference, frequency offsets also lead to attenuation and rotation of the sub-carrier phases.

4.2.1 Modelling frequency offset errors

In general, an offset in frequency is caused by a misalignment between receiver and transmitter local oscillator frequencies or due to Doppler shift. The Doppler frequency shift f_{dk} is proportional to the sub-carrier frequency f_k, angle of the velocity vector α and the relative speed between the transmitter and the receiver v_r. The Doppler shift can be expressed as:

$$f_{dk} = \frac{v_r f_k}{c} \cos(\alpha) \tag{4.1}$$

In Eq. (4.1) c denotes the speed of light and it is approximately equal to 3×10^8 m/s. The frequency of each sub-carrier can be calculated by taking the sum of main carrier frequency f_c and baseband sub-carrier frequency f_{sc}; i.e.:

$$f_k = f_c \pm f_{sc} \tag{4.2}$$

Using Eq. (4.1) and Eq. (4.2), the relative frequency offset Δ_f due to Doppler shift can be expressed as the ratio between the actual frequency offset and sub-carrier spacing; i.e.:

$$\Delta_f = \frac{f_{dk}}{f_c \pm f_{sc}} \cos(\alpha) = \frac{v_r}{c} \cos(\alpha) \tag{4.3}$$

The frequency offset can be modelled at the receiver by multiplying the received signal in the time domain with a complex exponential whose frequency component is equal to the frequency offset value. If we denote the transmitted signal by $S[n]$ and the received signal by $R[n]$, the relation between the two under the influence of a frequency offset Δ_f can be given as:

$$R[n] = S[n]e^{j2\pi\Delta_f n/N + \phi_0} + w[n] \tag{4.4}$$

In Eq. (4.4) Δ_f denotes the relative frequency offset due to local oscillator mismatch or due to Doppler shift or due to combination of both. N stands for the total number of sub-carriers, ϕ_0 is the initial phase and w denotes additive white Gaussian noise (AWGN). Without loss of generality, we assume for the moment that $w(n) = 0$ and $\phi_0 = 0$; i.e., $R[n] = S[n]e^{j2\pi\Delta_f n/N}$.

4.2.2 Frequency offset in OFDM

In OFDM the frequency offset prevents the perfect alignment of FFT bins with the peaks of the sinc pulses, i.e., sub-carriers. This is illustrated in Figure 4.1 where the sampling mismatch due frequency offset is shown.

The FFT output corresponding to the kth sub-carrier (under frequency offset Δ_f) can be expressed as:

$$\begin{aligned} \hat{a}_{k'} &= \frac{1}{N}\sum_{n=0}^{N-1} R[n]e^{-j2\pi\frac{k'n}{N}} \\ &= \frac{1}{N}\sum_{k=0}^{N-1} a_k \sum_{n=0}^{N-1} e^{j2\pi\frac{kn}{N}} e^{j2\pi\Delta_f\frac{n}{N}} e^{-j2\pi\frac{k'n}{N}} \\ &= \frac{1}{N}\sum_{k=0}^{N-1} a_k \sum_{n=0}^{N-1} e^{j2\pi\frac{(k-k'+\Delta_f)n}{N}} \end{aligned} \tag{4.5}$$

Using the geometric series properties, Eq. (4.5) can also be expressed as [10], [12]:

$$\hat{a}_{k'} = \frac{1}{N}\sum_{k=0}^{N-1} a_k \frac{\sin(\pi(k-k'+\Delta_f))}{\sin\left(\frac{\pi(k-k'+\Delta_f)}{N}\right)} e^{j\pi\left(\frac{N-1}{N}\right)(k-k'+\Delta_f)} \tag{4.6}$$

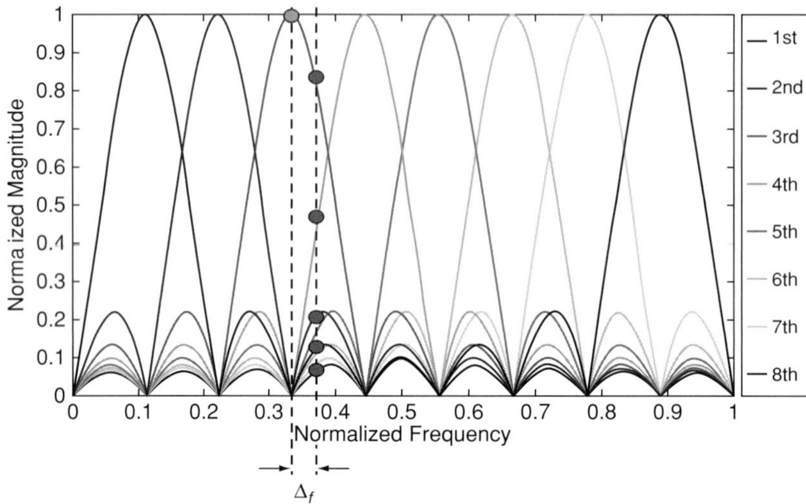

Figure 4.1 Illustration of erroneous sampling in OFDM due to carrier frequency offset.

We can split Eq. (4.6) into two distinct parts:

$$\hat{a}_{k'} = \underbrace{a_{k'} \frac{\sin(\pi \Delta_f)}{N \sin\left(\frac{\pi \Delta_f}{N}\right)} e^{j\pi\left(\frac{N-1}{N}\right)\Delta_f}}_{\text{Useful Signal (Attenuated, Phase Shifted)}} + \underbrace{\frac{1}{N} \sum_{k=0;k \neq k'}^{N-1} a_k \frac{\sin(\pi(k-k'+\Delta_f))}{\sin\left(\frac{\pi(k-k'+\Delta_f)}{N}\right)} e^{j\pi\left(\frac{N-1}{N}\right)(k-k'+\Delta_f)}}_{\text{Intercarrier Interference (ICI)}} \tag{4.7}$$

The first component of Eq. (4.7) stands for useful demodulated signal, which has been attenuated and phase shifted due to frequency offset. Since the attenuation term is not dependent on the index of the carrier, all the sub-carriers experience the same degree of attenuation [26]. The second part of Eq. (4.7) contains the inter-carrier interference (ICI) term, which represents the deleterious impact of all other sub-carriers on the decision making of data contained in the carrier of interest. The carrier frequency offset (CFO) does not influence the amplitude of the OFDM signal and therefore the total received power is not altered. Furthermore, the total ICI power due to CFO is also not affected by the number of OFDM carriers [26].

4.2.3 Frequency offset in WPM

The detected data for the kth sub-carrier and the uth symbol at the WPM receiver under a loss of frequency synchronization can be expressed as:

$$\begin{aligned}\hat{a}_{u',k'} &= \sum_n R[n]\xi_l^{k'}[u'N-n] = \sum_n \sum_u \sum_{k=0}^{N-1} a_{u,k}\xi_l^k[n-uN]e^{j2\pi\Delta_f\left(\frac{n}{N}\right)}\xi_l^{k'}[u'N-n] \\ &= \sum_u \sum_{k=0}^{N-1} a_{u,k}\left(\sum_n \xi_l^k[n-uN]e^{j2\pi\Delta_f\left(\frac{n}{N}\right)}\xi_l^{k'}[u'N-n]\right)\end{aligned} \tag{4.8}$$

Table 4.1 Simulation setup for study of frequency offset effects

	WPM	OFDM
Number of sub-carriers	128	128
Number of multi-carrier symbols per frame	100	100
Modulation	QPSK	QPSK
Channel	AWGN	AWGN
Oversampling factor	1	1
Guard band	–	–
Guard interval	–	–
Carrier frequency f_c	0 (base band)	0 (base band)
Frequency offset	$\Delta_f = 5$–10%	$\Delta_f = 5$–10%
Phase noise	–	–
Time offset	–	–

Defining the cross-waveform function $\Omega[\Delta_f]$ as:

$$\Omega_{k,k'}^{u,u'}[\Delta_f] = \sum_n e^{j2\pi\Delta_f(\frac{n}{N})}\xi_l^k[n-uN]\xi_l^{k'}[u'N-n], \tag{4.9}$$

the demodulated information bit of the kth sub-carrier and the uth WPM symbol corrupted by the interference due to loss of orthogonality can be expressed as:

$$\hat{a}_{u',k'} = \underbrace{a_{u',k'}\Omega_{k'k'}^{u',u'}[\Delta_f]}_{\text{Desired Alphabet}} + \underbrace{\sum_{u;u\neq u'} a_{u,k'}\Omega_{k',k'}^{u,u'}[\Delta_f]}_{\text{ISI}} + \underbrace{\sum_u \sum_{k=0;k\neq k'}^{N-1} a_{u,k}\Omega_{k,k'}^{u,u'}[\Delta_f]}_{\text{Inter Symbol-ICI (IS-ICI)}} \tag{4.10}$$

In Eq. (4.10) the first term stands for the attenuated and rotated version of the useful data. The second term gives the ISI due to symbols transmitted on the same sub-channel and the third term denotes ICI measured over the whole frame. Indeed, the orthogonality condition $\sum_n \xi_l^k[n-uN]\xi_l^{k'}[u'N-n] = \delta[u-u']\delta[n-n']$ is lost only because of the frequency offset Δ_f.

4.2.4 Numerical results for frequency offset

In this section we investigate the performance of WPM under frequency offset by means of computer simulations. The WPM transceiver is simulated with different wavelet families and compared to the well-known OFDM. To simplify the analysis, the channel is taken to be additive white Gaussian noise (AWGN) and no other distortions except frequency offset are introduced. QPSK is the modulation of choice and the frame size is set to 100 multi-carrier symbols, each consisting of 128 sub-carriers. Furthermore, the simulated system has no error estimation or correction capabilities nor are guard intervals or guard bands used. Any deviation from these specifications will be explicitly stated. The parameters of the simulation set-up are provided in Table 4.1.

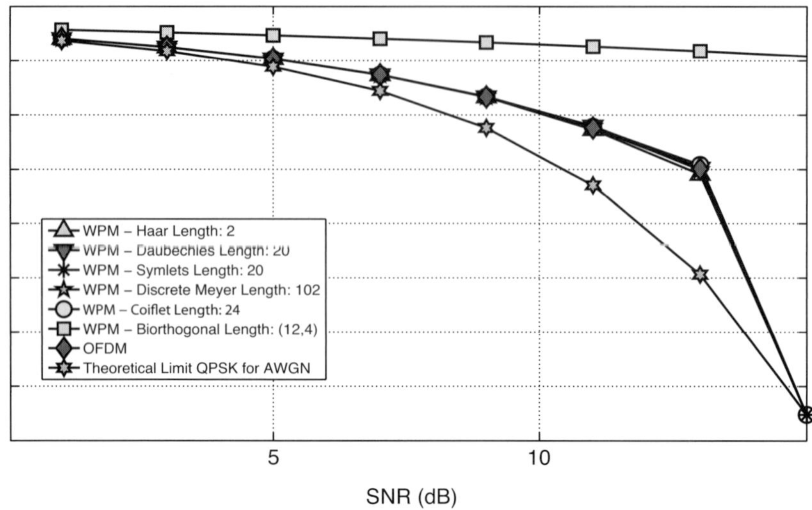

Figure 4.2 BER for WPM for different wavelets and OFDM under a relative frequency offset of 5%.

4.2.4.1 Performance under frequency offset error

Figure 4.2 shows the bit error rate (BER) plots of OFDM and WPM transceivers with relative frequency offset of 5% with regard to $^1/_T$ spacing. BER curves of different wavelets and OFDM show similar performance but due to frequency offset they all lie far from the theoretical curve. The biorthogonal wavelet is the exception with a very poor performance compared to the other systems. This is due to the fact that the biorthogonal wavelets do not fulfil the orthogonality condition.

In Figure 4.3 the BER is shown for different values of relative frequency offset varying from 0 to 40% for a constant signal-to-noise ratio (SNR) value of 16 dB. Again we can see that the performances of majority of the wavelets are very similar to that of OFDM. The biorthogonal wavelet has obviously a poor performance, while the Haar wavelet slightly outperforms other wavelets and OFDM. The results also make clear the sensitivity of both WPM and OFDM systems to frequency offset.

4.2.4.2 Influence of number of sub-carriers

The results of the investigation on the influence of the number of sub-carriers on the BER performance of the WPM systems under frequency offset is depicted in Figure 4.4. All WPM transceivers are now simulated with the same wavelet but with different number of sub-carriers. We arbitrarily chose the Daubechies wavelet with 20 coefficients. Furthermore, the relative frequency offset is set to 10% and again we use AWGN channel.

The degradation of WPM performance in the presence of frequency offset is dependent on the number of sub-carriers. This dependency is straightforward when the absolute frequency offset is fixed [1]. As the number of sub-carriers in a given bandwidth increases, the spacing between the sub-carriers decreases and hence the relative frequency offset increases. The results of these studies are plotted in Figure 4.4. In the

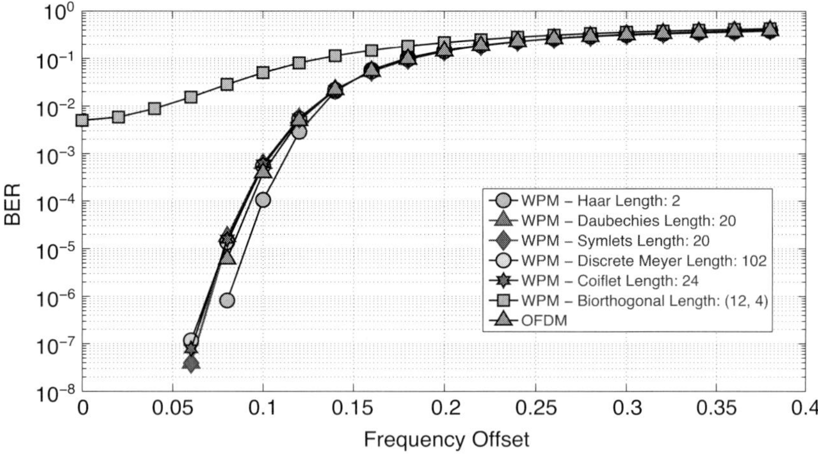

Figure 4.3 BER vs. relative frequency offset for WPM and OFDM in AWGN channel (SNR = 16 dB).

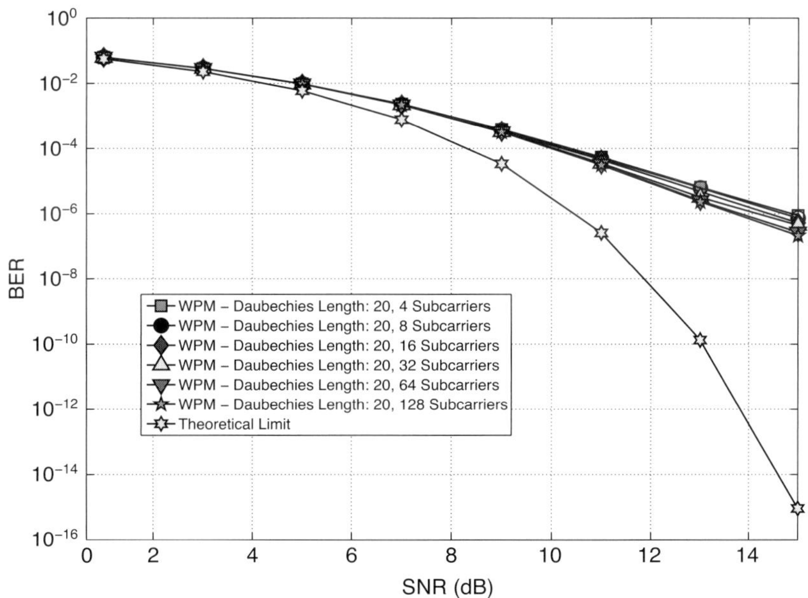

Figure 4.4 BER for WPM with different number of sub-carriers. Relative frequency offset = 10%. Wavelet of choice: Daubechies 10.

scenario considered while plotting Figure 4.4, the relative frequency offset with respect to inter-carrier spacing is kept constant. The WPM configurations with larger number of sub-carriers are more susceptible to the frequency offset. However, beyond a point this sensitivity saturates even with increasing number of sub-carriers. For example, from

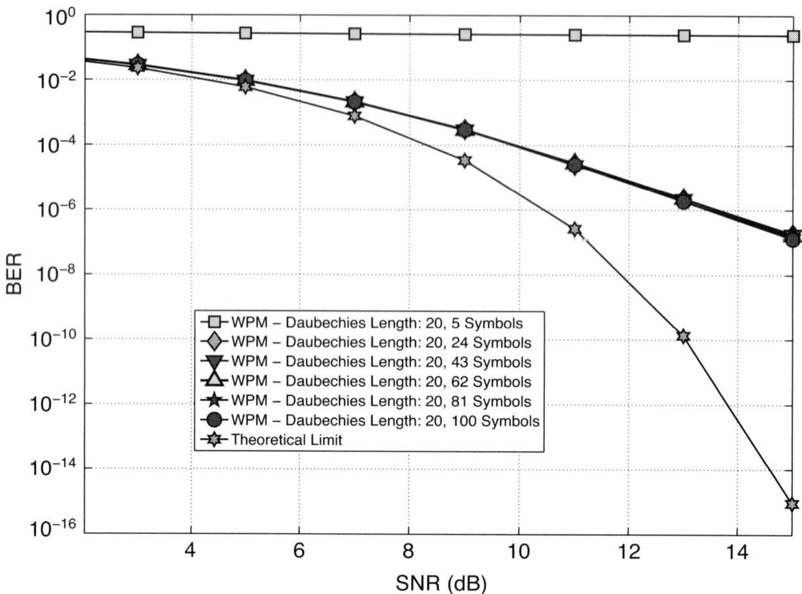

Figure 4.5 BER for WPM with different number of multi-carrier symbols/frame and relative frequency offset of 10%. Wavelet family: Daubechies.

Figure 4.4, we can also see that the performances of WPM with 64 and 128 sub-carriers are almost identical for a given relative frequency offset.

4.2.4.3 Influence of WPM frame size

Frequency offset in WPM not only leads to ICI within one symbol but also across the whole frame. Therefore, it is important to see the effect of the frame size in combination with the frequency offset. These results are illustrated in Figure 4.5. The plots show that the number of symbols in a frame does not affect the performance of WPM in the presence of frequency offset.

4.2.4.4 Influence of wavelet filter length

The influence of the filter's length in combination with the frequency offset on the BER is illustrated in Figure 4.6. This simulation is performed for the AWGN channel and the relative frequency offset of 10%. We again choose the Daubechies wavelet but now we alter the number of filter coefficients and fix the number of sub-carriers to 128. In the case of Daubechies wavelet the number of wavelet zero moments is indisputably related to the length of the filter, and hence for each doubling of filter coefficients we also double the number of wavelet zero moments.

The BER curves shown in Figure 4.6 are all superimposed one over another, suggesting that the filter's length and number of wavelets' zero moments have no tangible influence on the system performance in the case of frequency offset.

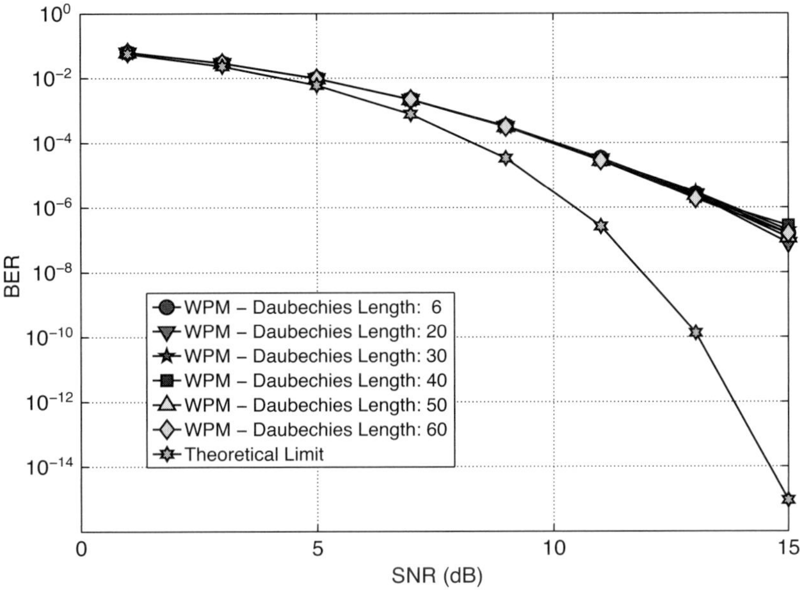

Figure 4.6 BER for WPM using Daubechies wavelets of different lengths and relative frequency offset of 10%.

4.2.4.5 Constellation plots

The effect of frequency misalignment between transmitter and receiver on the constellation points is depicted in Figure 4.7, for the relative frequency offset of 5%. In order to highlight the effect of frequency offset, we assumed an ideal channel without any other infarction or noise barring a loss in frequency synchronization. The main effect of the frequency offset is the scattering of the constellation points around reference positions due to interference. Other consequences are the anti-clockwise rotation of all constellation points and a marginal attenuation.

4.2.4.6 Dispersion of sub-carrier energy

The last set of figures in this section shows the dispersion of the sub-carriers energy due to frequency offset (Figure 4.8 for WPM and Figure 4.9 for OFDM). For clarity, we limited the number of sub-carriers to 16 and the frame size to 30 multi-carrier symbols. The channel is assumed to be ideal and all disturbances in the transmission are solely due to the frequency offset. Figures 4.8 and 4.9 were obtained by transmitting a single non-zero pilot sub-carrier with all other sub-carriers in the frame set to zero.

In an ideal situation, without any frequency offset, the only sub-carrier with non-zero value will be the pilot sub-carrier, regardless of which system we use: WPM or OFDM. However, the frequency offset results in loss of orthogonality and sub-carriers begin to interfere with one another. In OFDM the effect of frequency offset is to introduce inter-carrier interference. This infarction is confined within a single multi-carrier symbol and

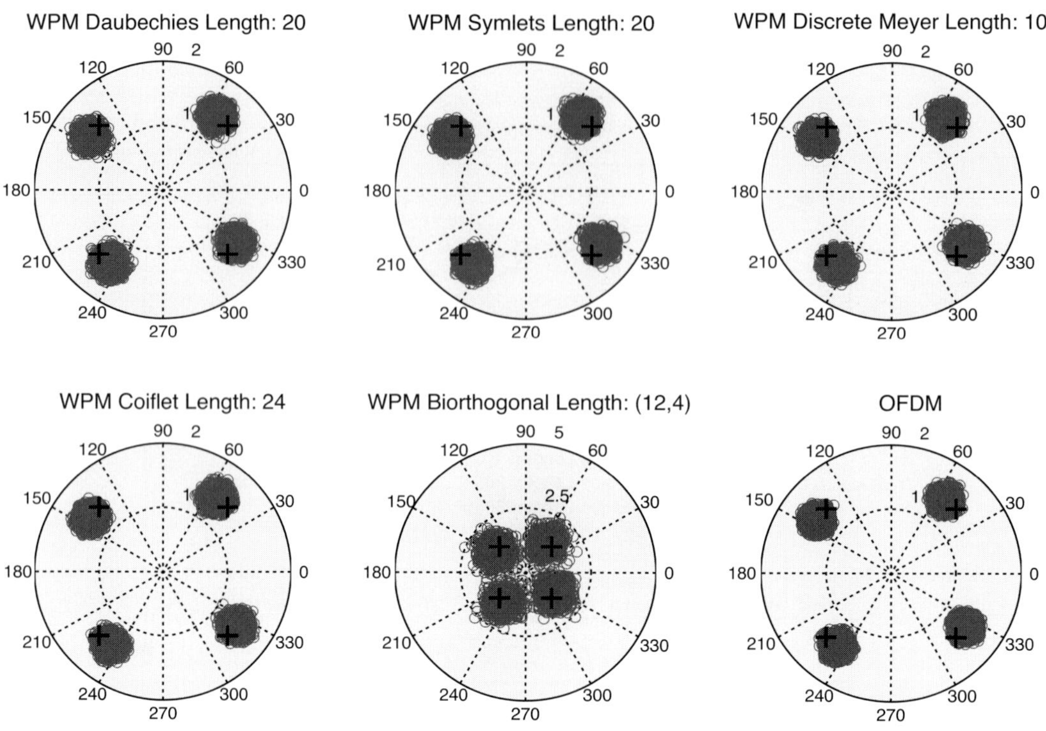

Figure 4.7 Constellation points in the presence of relative frequency offset of 5%.

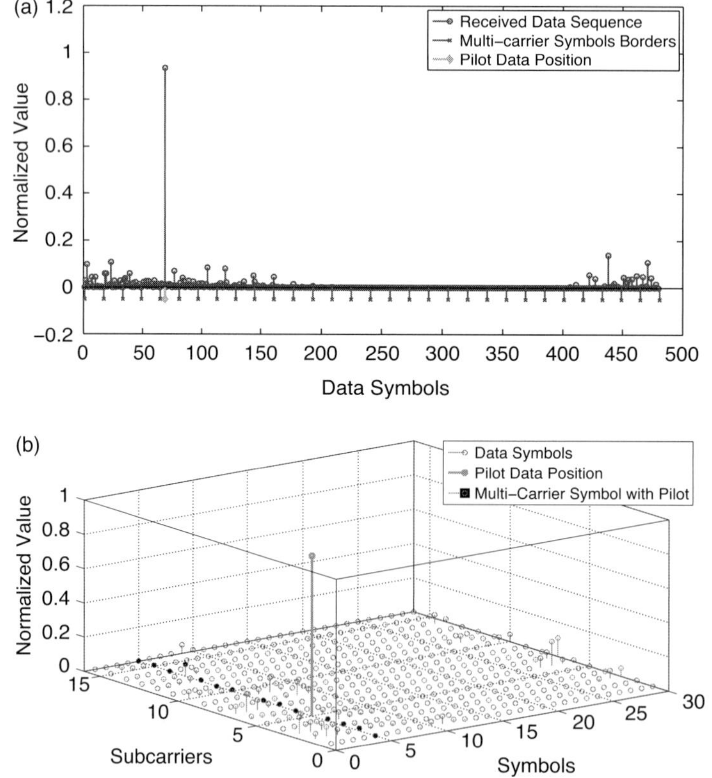

Figure 4.8 Spectral energy in a frame of received WPM signal affected by a frequency offset. (a) 2D and (b) 3D. Wavelet used: Daubechies 10.

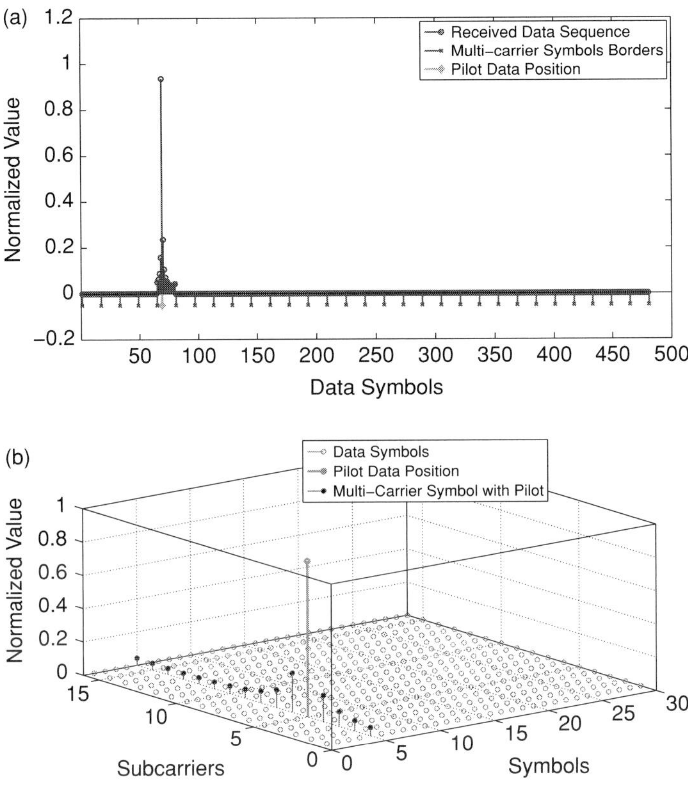

Figure 4.9 Spectral energy in a frame of received OFDM signal affected by a frequency offset. (a) 2D and (b) 3D.

other OFDM symbols are not affected. The WPM, on the other hand, has overlapping symbols and an offset in frequency results in both ICI and inter-symbol-ICI.

In Figure 4.8 we therefore observe that the energy of the pilot sub-carrier located in the 5th sub-carrier and 5th symbol is spread almost across the whole frame. This is in agreement with the theoretical derivation carried in Sections 4.2.2 and 4.2.3.

4.3 Phase noise in multi-carrier modulation

The ideal local oscillator would have a single carrier with constant amplitude and frequency. However, the outputs of practical local oscillators are degraded due to factors such as thermal noise [27], causing the oscillator's central frequency to fluctuate a bit. This uncertainty in the actual frequency or the phase of the signal is referred to as phase noise. Multi-carrier transmission is very vulnerable to phase noise since phase noise can cause the loss of orthogonality between sub-carriers.

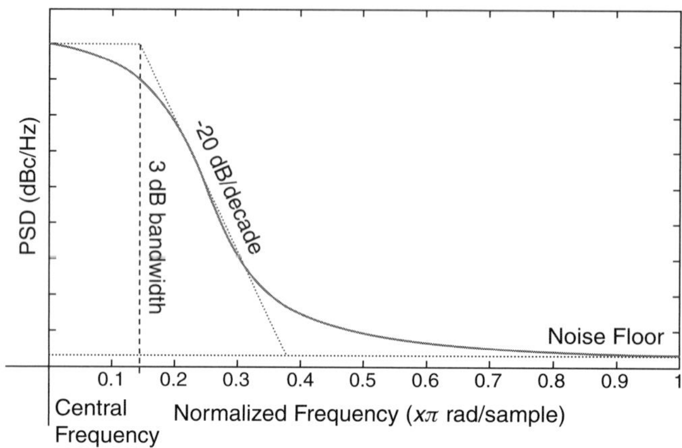

Figure 4.10 Single side-band PSD of the oscillator process.

4.3.1 Modelling the phase noise

The perturbance caused by the phase noise on multi-carrier transmission can be divided into two components:

- Common phase error (CPE): Attenuates and rotates all constellation symbols by the same angle.
- Interference: Contribution of all other sub-carriers.

Phase noise can be represented as a parasitic phase modulation of the oscillator's signal. In the literature there are different models used for the phase noise. The majority of these models are described in terms of power spectral density (PSD). In the ideal case the PSD of the local oscillator would be a single pulse (delta function) at the central frequency. Due to imperfections of the oscillator (fluctuations), the PSD of the practical oscillator is distributed over a wider frequency band with the highest concentration around oscillator's central frequency. The single side-band PSD of free running oscillator can be estimated by the Lorentzian function [28], like the one illustrated in Figure 4.10.

In this work we model the phase noise as a zero mean white Gaussian process ϕ_w with finite variance σ_w^2 [5]. The autocorrelation function of the phase noise is given by:

$$R_{\phi_w}[m] = \sigma_w^2 \delta[m] \tag{4.11}$$

The power spectral density of the phase noise can be expressed as:

$$S_{\phi_w}(f) = \sum_{m=-\infty}^{\infty} R_{\phi_w}[m] e^{-j2\pi f m} \tag{4.12}$$

In order to get the desired phase-noise bandwidth, we perform low-pass filtering with filter F_ϕ. The PSD in Eq. (4.12) now becomes:

$$S_{\phi b}(f) = S_{\phi_w}(f) \left| F_\phi(f) \right|^2 \tag{4.13}$$

By changing the corner frequency $f_{c\phi}$ of the filter used, we can adjust the phase-noise bandwidth. A low value of corner frequency results in a narrow bandwidth while higher values spread the phase noise.

In the last stage of the model we add phase noise floor to the signal. Similarly to the main phase noise contribution, the phase noise floor is also modelled as a zero mean Gaussian process with finite variance σ_{wn}^2, which is relatively low compared to σ_w^2. The phase noise floor is not correlated so that it spans the whole available bandwidth and has flat PSD.

The total phase noise Δ_ϕ can now be expressed as a sum of bandwidth-limited main noise contribution ϕ_b and phase noise floor ϕ_{wn} as:

$$\Delta_\phi[n] = \phi_b[n] + \phi_{wn}[n] \tag{4.14}$$

Using the phase noise model given in Eqs. (4.11)–(4.14), we can write the received signal $R[n]$ that has been affected by phase noise and AWGN channel as:

$$R[n] = S[n]e^{j\Delta_\phi[n]} + w[n] \tag{4.15}$$

Without loss of generality, we assume for the moment that $w[n] = 0$; i.e., $R[n] = S[n]e^{j\Delta_\phi[n]}$.

4.3.2 Phase noise in OFDM

When an OFDM transceiver experiences some phase noise, we can express the demodulated signal at the receiver's output as:

$$\begin{aligned} \hat{a}_{k'} &= \frac{1}{N}\sum_{n=0}^{N-1} R[n]e^{-j2\pi\frac{k'}{N}n} = \frac{1}{N}\sum_{k=0}^{N-1} a_k \sum_{n=0}^{N-1} e^{j2\pi\frac{k}{N}n} e^{j\Delta_\phi[n]} e^{-j2\pi\frac{k'}{N}n} \\ &= \frac{1}{N}\sum_{k=0}^{N-1} a_k \sum_{n=0}^{N-1} e^{j\Delta_\phi[n]} e^{j2\pi\frac{[k-k']n}{N}} \end{aligned} \tag{4.16}$$

We can simplify the analysis for phase noise by splitting the demultiplexed signal into a useful part and a disturbance part. In order to do this we will assume that phase noise is sufficiently small that it can be approximated by [5]:

$$e^{j\Delta_\phi[n]} \approx 1 + j\Delta_\phi[n] \tag{4.17}$$

Using approximation Eq. (4.17), we can express the demodulated OFDM signal Eq. (4.16) for the kth carrier as:

$$\begin{aligned} \hat{a}_{k'} &\approx \frac{1}{N}\sum_{k=0}^{N-1} a_k \sum_{n=0}^{N-1} e^{j2\pi\frac{[k-k']n}{N}} + \frac{j}{N}\sum_{k=0}^{N-1} a_k \sum_{n=0}^{N-1} \Delta_\phi[n] e^{j2\pi\frac{[k-k']n}{N}} \\ &\approx a_{k'} + \frac{j}{N}\sum_{k=0}^{N-1} a_k \sum_{n=0}^{N-1} \Delta_\phi[n] e^{j2\pi\frac{[k-k']n}{N}} = a_{k'} + I_{\Delta_\phi}[k] \end{aligned} \tag{4.18}$$

The first component of Eq. (4.18) stands for a correctly demodulated symbol and the second term $I_{\Delta\phi}$ stands for a disturbance that is added to each sub-carrier. Two distinct scenarios are possible with the phase noise:

1. If $k' = k$: *common phase error (CPE)*

The disturbance term from Eq. (4.18) can now be written as:

$$\begin{aligned} I_{\Delta_\phi}[k'] &= \frac{j}{N}\sum_{k=0}^{N-1} a_k \sum_{n=0}^{N-1} \Delta_\phi[n] \\ &= j\Phi a_{k'} \end{aligned} \tag{4.19}$$

The error, given in Eq. (4.19), causes the constellation points to be rotated by an angle Φ. This angle is common for all sub-carriers so that all constellation points will be rotated by the same angle. Here, the rotation angle Φ is defined by the average phase noise given as:

$$\Phi = \frac{1}{N}\sum_{n=0}^{N-1} \phi[n] \tag{4.20}$$

The common phase error (CPE) is only dependent on low frequencies of the phase noise spectrum up to the frequency of the inter-carrier spacing.

2. If $k' \neq k$: *inter-carrier interference (ICI)*

The disturbance term from Eq. (4.18) can now be written as:

$$\underset{k\neq k'}{I_{\Delta_\phi}[k]} = \frac{j}{N}\sum_{k=0;k\neq k'}^{N-1} a_k \sum_{n=0}^{N-1} \Delta_\phi[n] e^{j2\pi\frac{[k-k']n}{N}} \tag{4.21}$$

The error in Eq. (4.21) consists of contribution from all other sub-carriers in an OFDM symbol, and it is known as ICI. The magnitude of ICI as a result of phase noise is dependent only at the phase-noise components that have high frequencies. In general, the phase noise that causes ICI contains frequencies that are larger than the inter-carrier spacing frequency.

4.3.3 Phase noise in WPM

The detected data at the WPM receiver in presence of the phase noise for the kth carrier and uth symbol can be written as:

$$\begin{aligned} \hat{a}_{u',k'} &= \sum_n R[n]\xi_l^{k'}[u'N-n] \\ &= \sum_n \sum_u \sum_{k=0}^{N-1} a_{u,k}\xi_l^k[n-uN]e^{j\Delta_\phi[n]}\xi_l^{k'}[u'N-n] \\ &= \sum_u \sum_{k=0}^{N-1} a_{u,k}\left(\sum_n \xi_l^k[n-uN]e^{j\Delta_\phi[n]}\xi_l^{k'}[u'N-n]\right) \end{aligned} \tag{4.22}$$

Defining the cross-waveform function $\Omega\,[\Delta_\phi]$ as:

$$\Omega_{k,k'}^{u,u'}[\Delta_\phi] = \sum_n \xi_l^k[n - uN]\Delta_\phi[n]\xi_l^{k'}[u'N - n] \tag{4.23}$$

and assuming that the phase noise is sufficiently small that it can be approximated as [5]:

$$e^{j\Delta_\phi[n]} \approx 1 + j\Delta_\phi[n] \tag{4.24}$$

we can rewrite Eq. (4.22) as:

$$\begin{aligned}
\hat{a}_{u',k'} &\approx \sum_u \sum_{k=0}^{N-1} a_{u,k}\left(\sum_n \xi_l^k[n - uN]\xi_l^{k'}[u'N - n]\right) \\
&\quad + j\sum_u \sum_{k=0}^{N-1} a_{u,k}\left(\sum_n \xi_l^k[n - uN]\Delta_\phi[n]\xi_l^{k'}[u'N - n]\right) \\
&\approx \underbrace{a_{u',k'}}_{\substack{\{u=u',k=k'\}\\ \text{Useful Data}}} + \underbrace{I_{\Delta_\phi}[u,k]}_{\text{Interference Term}}
\end{aligned} \tag{4.25}$$

The first component of Eq. (4.25) stands for a correctly demodulated symbol and the second term $I_{\Delta_\phi}(u, k)$ stands for a disturbance due to the phase offset. Two distinct scenarious arise out of the error term.

1. If $k' = k$ and $u' = u$: common phase error (CPE)

The disturbance term from Eq. (4.25) can now be written as

$$I_{\Delta_\phi}[u', k'] = \underbrace{j a_{u',k'}\Omega_{k',k'}^{u',u'}[\Delta_\phi[n]]}_{\substack{\{u=u',k=k'\}\\ \text{Useful Data biased by Common Phase Error}}} \tag{4.26}$$

that describes the rotation of constellation points by an angle that is common for all sub-carriers. The rotation angle is dependent on the average value of phase noise sequence.

2. If $k' \neq k$ and/or $u \neq u'$: inter-carrier interference (ICI) and inter-symbol–inter-carrier interference (IC–ISI)

The interference term from Eq. (4.22) can now be written as

$$I_{\Delta_\phi}[u, k] = \underbrace{j\sum_{k=0;k\neq k'}^{N-1} a_{u',k}\Omega_{k,k'}^{u',u'}[\Delta_\phi] +}_{\substack{\{u=u',k\neq k'\}\\ \text{Inter Carrier Interference}}} \underbrace{j\sum_{u;u\neq u'}\sum_{k=0}^{N-1} a_{u,k}\Omega_{k,k'}^{u,u'}[\Delta_\phi]}_{\substack{\{u\neq u'\}\\ \text{Inter Symbol-Inter Carrier Interference}}} \tag{4.27}$$

The first term stands for the inter-carrier interference and the second for inter-symbol–inter-carrier interference. In contrast to the OFDM in an AWGN channel, the phase noise in WPM raises the ICI and ISI levels. This is due to the overlapping nature of the wavelet transform.

The demodulated data hence consist of the estimate of the useful data and the interference terms:

Table 4.2 Simulation setup for studying the impact of phase noise

	WPM	OFDM
Number of sub-carriers	128	128
Number of multi-carrier symbols per frame	100	100
Modulation	QPSK	QPSK
Channel	AWGN	AWGN
Oversampling factor	1	1
Guard band	–	–
Guard interval	–	–
Frequency offset	–	–
Phase noise	$\sigma_w^2 = -10$ dBc, $\sigma_{wn}^2 = -20$ dBc, $f_{c\phi} = 0.1$	$\sigma_w^2 = -10$ dBc, $\sigma_{wn}^2 = -20$ dBc, $f_{c\phi} = 0.1$
Time offset	–	–

- common phase error ($u = u'$ and $k' = k$);
- inter-carrier interference ($u = u'$ and $k' \neq k$);
- inter-symbol–inter-carrier interference ($u \neq u'$, $k' = k$ and $u \neq u'$, $k' \neq k$).

Different frequency components of the phase noise have different impacts on the CPE and ICI/ISI terms. If the phase-noise bandwidth is concentrated near the central frequency the CPE term will dominate, but when the phase-noise bandwidth is somewhat more spread the ICI/ISI term will soon take over. Using already defined model for phase noise Eqs. (4.11)–(4.14), we can control the phase-noise bandwidth by the corner frequency of the filter.

4.3.4 Numerical results for phase noise

The performance degradation associated with phase noise has been evaluated by the computer simulations using an almost identical set-up as for the frequency offset. More details on the set-up can be found in Section 4.2.4. An overview of simulation parameters is given in Table 4.2.

4.3.4.1 Phase-noise characteristics

The different effects of the phase noise on the WPM and OFDM are best illustrated by the constellation points diagram and the PSD of the phase noise (Figure 4.11 for narrowband phase noise and Figure 4.12 for wideband phase noise). Figure 4.11a shows the PSD of the phase noise which has a relatively low corner frequency. The dominant effect of such phase noise is the common phase error, which results in the rotation of all constellation points. The PSD of the phase noise with a relatively high corner frequency is illustrated in Figure 4.12a. Now, the rotation behaviour is not more visible but the interference between sub-carriers is much more pronounced.

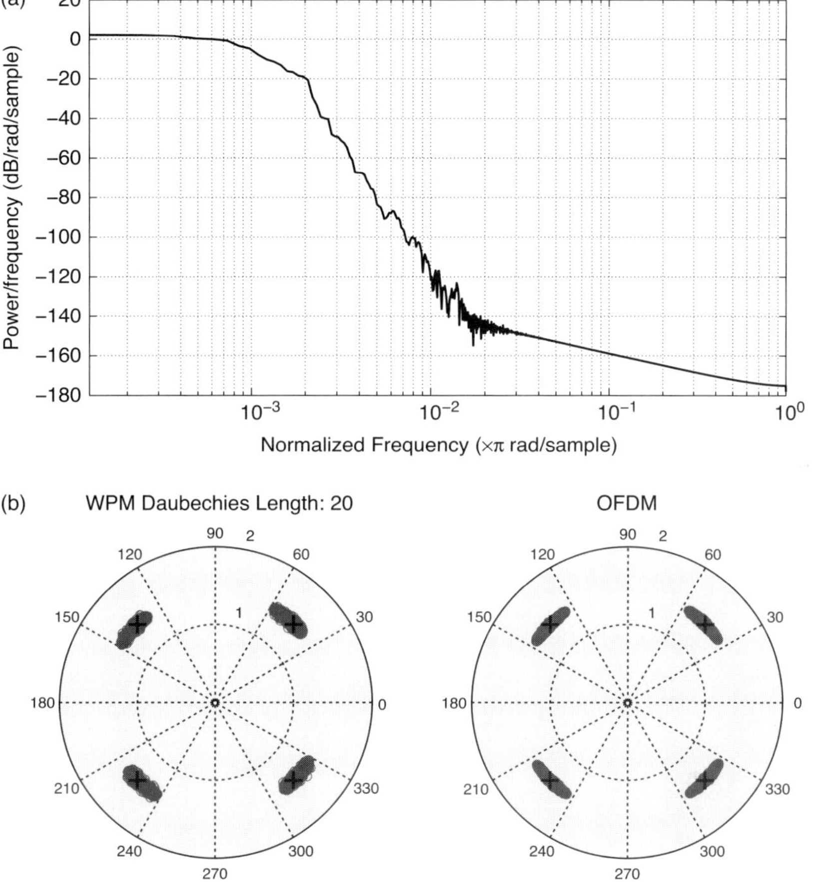

Figure 4.11 Phase noise (narrow band); (a): PSD, (b): WPM and OFDM constellation points.

Both effects of the phase noise are important and depending on the system one or the other can be the limiting factor for the system performance. In the literature there are many adequate correction approaches available for the CPE [15]–[16] but the estimation and correction of the interference is much harder to realize. Therefore, we will limit ourselves to the performance analysis of the WPM and OFDM in the presence of phase noise that causes interference. In order to achieve this, we set the phase-noise bandwidth to 10% of the total available bandwidth and the variance to –10 dBc (decibel relative to the carrier). The PSD of the phase noise will look similar to one illustrated in Figure 4.10.

4.3.4.2 Performance under frequency offset error

Figure 4.13 shows the bit error rate (BER) of WPM and OFDM in the presence of phase noise. The illustrated behaviours of BER curves are similar to each other with the exception of the biorthogonal wavelet. The poor performance of the

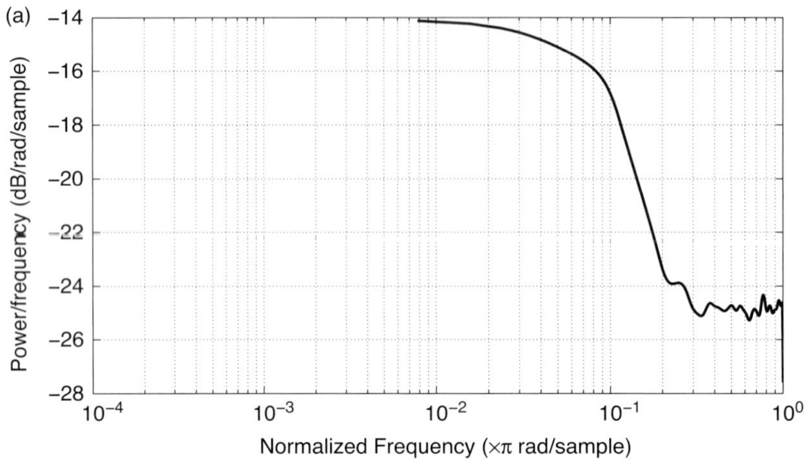

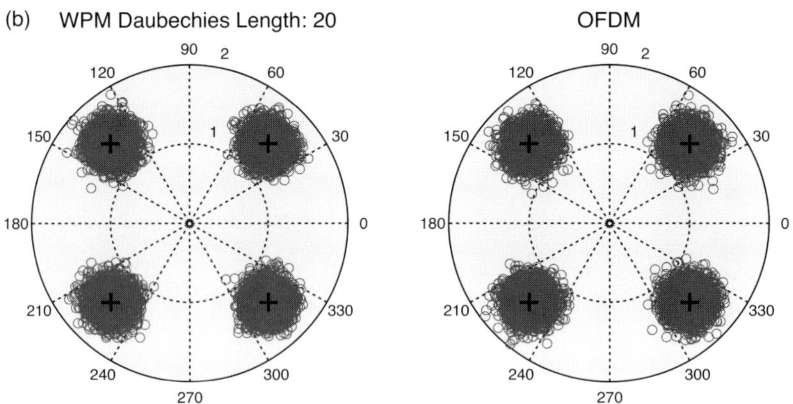

Figure 4.12 Phase noise (wide band); (a): PSD, (b): WPM and OFDM constellation points.

biorthogonal wavelet, as already mentioned, is due to unfulfiled perfect reconstruction constraint.

Figure 4.14 illustrates the effect of the phase-noise variance on the BER. This figure is obtained using an AWGN channel with 16 dB SNR, while phase-noise variance is varied from –10 to 20 dBc with a step size of 5 dBc. It is natural that the phase-noise variance and the performance degradation are closely related. The sensitivity of WPM and OFDM to the variance of the phase noise is confirmed by Figure 4.14.

4.3.4.3 Influence of number of sub-carriers and WPM frame size

Figures 4.15 and 4.16, respectively, show the performance of the WPM with phase noise when the number of sub-carriers and symbols in the frame is altered. The simulation results have not shown any essential connection between performance degradation and the number of sub-carriers nor the number of symbols per frame. The results would

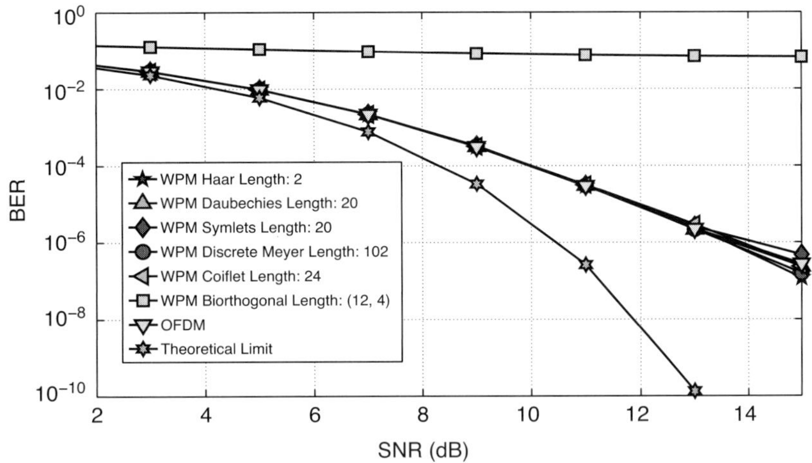

Figure 4.13 BER for WPM with different wavelets and OFDM under phase noise with relative bandwidth of 10% and variance of –10 dBc.

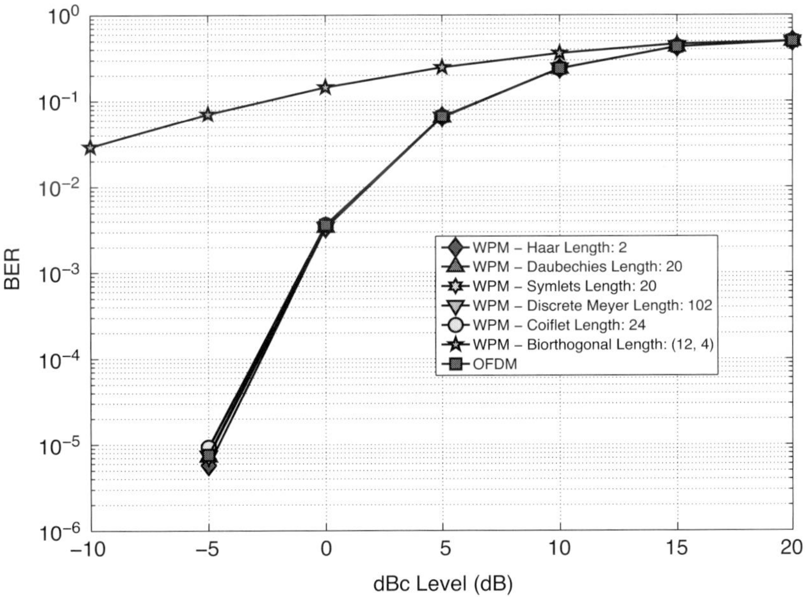

Figure 4.14 BER vs. phase-noise variance for WPM and OFDM in AWGN channel (SNR = 16 dB).

be different if we set the corner frequency to a smaller value, because the inter-carrier spacing depends on the number of sub-carriers, for low numbers of sub-carriers dominant CPE term will dominate, while for high numbers of sub-carriers the interference will be the major term [5], [29].

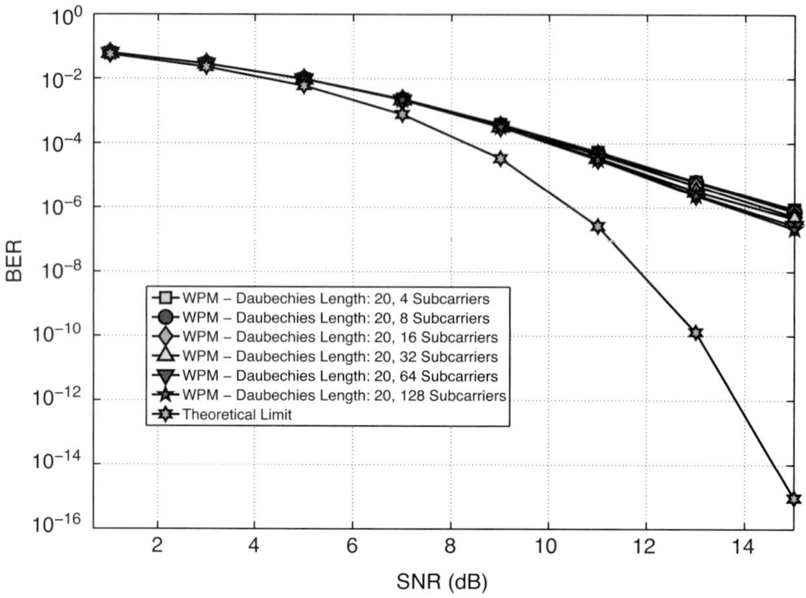

Figure 4.15 BER for WPM with phase noise for different numbers of sub-carriers.

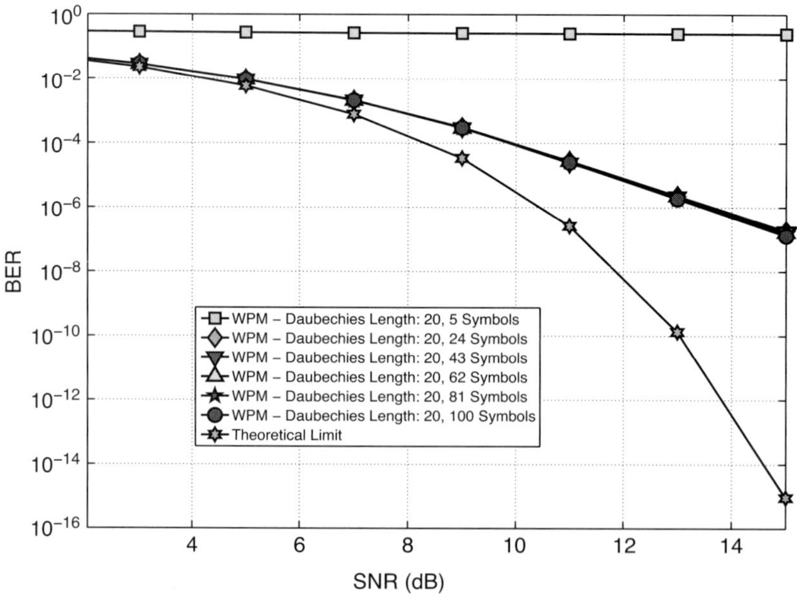

Figure 4.16 BER for WPM with phase noise for different numbers of symbols/frame.

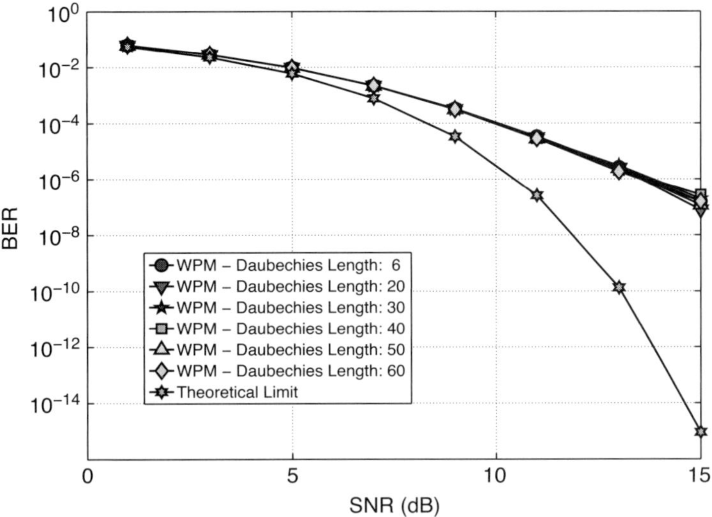

Figure 4.17 BER for WPM using Daubechies wavelets of different lengths under influence of phase noise.

4.3.4.4 Influence of wavelet filter length

Figure 4.17 illustrates the influence of the filter length and number of zero wavelet moments in combination with the phase noise on the BER. As with frequency offset, there are no noticeable influences of the filter's length and number of wavelets' zero moments on the system performance when operating under a phase noise.

4.3.4.5 Constellation plots

For completeness of the analysis we show in Figure 4.18 the effect of the phase noise on the constellation points, but now for all discussed wavelets and OFDM. The clearly visible scattering of the constellation points around the reference positions is caused by the phase noise, as the channel is assumed to be ideal and no other disturbances were introduced.

The spreading of sub-carrier energy due to phase noise is illustrated in Figures 4.19a and b for OFDM and Figures 4.20a and b for WPM. Phase noise results in loss of orthogonality and causes sub-carriers to interfere with each other. In OFDM, interference due to phase noise is limited to within a symbol resulting in an ICI. However, due to overlap of symbols in WPM, the phase noise causes ICI from other symbols resulting in inter-symbol–inter-carrier interference (IS–ICI). This is illustrated in Figure 4.20(b) where the dispersion of energy of a *pilot* sub-carrier is shown to extend to the entire frame. In the example presented we consider a pilot sub-carrier located at the 5th sub-carrier of the 5th symbol. This dispersion in energy is in agreement with the theoretical derivations carried out in Sections 4.3.2 and 4.3.3.

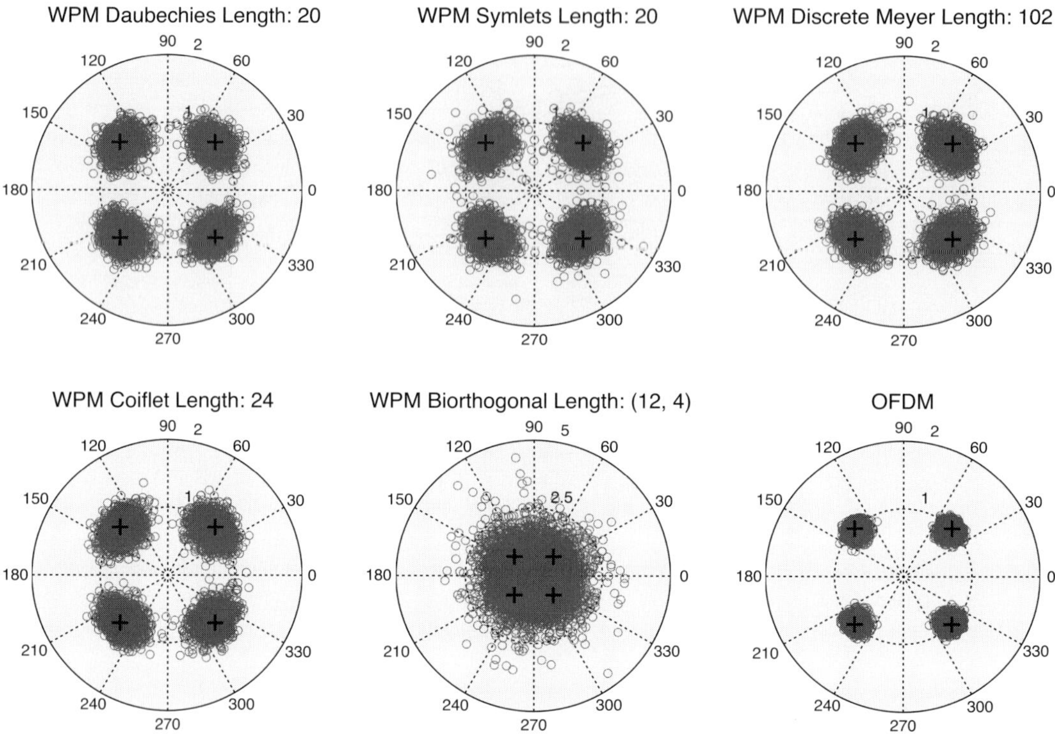

Figure 4.18 Constellation points in the presence of phase noise.

4.4 Time offset in multi-carrier modulation

One more drawback of multi-carrier transmissions is their vulnerability to time synchronization errors that occur when multi-carrier symbols are not perfectly aligned at the receiver. Because of the time offset, samples outside a WPM or OFDM symbol get erroneously selected, while useful samples at the beginning or at the end of that particular symbol get discarded.

4.4.1 Modelling time offset errors

The time synchronization error is modelled by shifting the received data samples by a time offset value Δ_t[2] to the left or right, depending on the sign of the Δ_t [18]. If we assume that the transmitted signal is given by $S[n]$, the received signal $R[n]$ in the presence of time synchronization error can be expressed as:

$$R[n \pm \Delta_t] = S[n] + w[n] \tag{4.28}$$

[2] The variations in time Δ_t are usually modelled as a stochastic process.

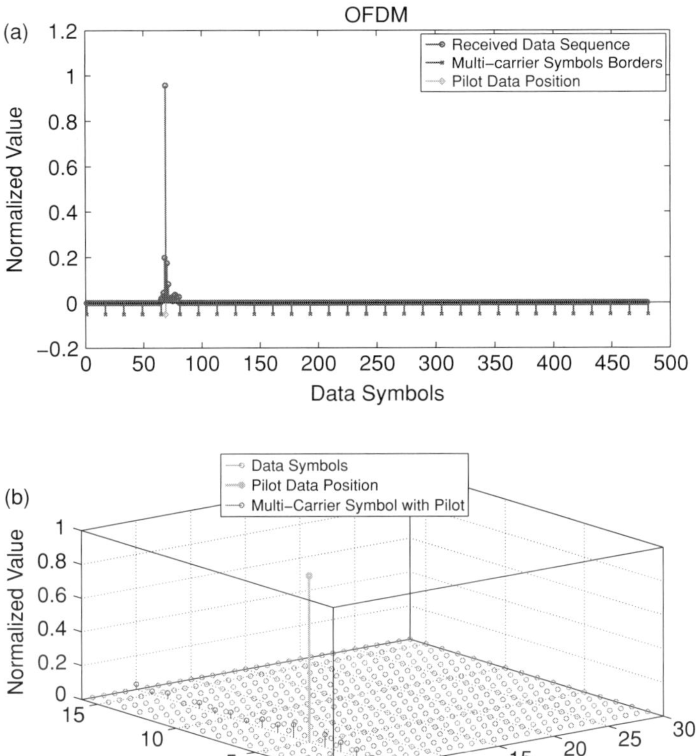

Figure 4.19 Spectral energy in a frame of received OFDM signal affected by phase noise. (a) 2D and (b) 3D view.

Without loss of generality, we assume $w[n] = 0$, then, $R[n \pm \Delta_t] = S[n]$. Time offset degrades the performances of multi-carrier transceivers by introducing inter-symbol interference (ISI).

WPM and OFDM share many similarities as both are orthogonal multi-carrier systems but with regard to timing error the behaviours are vastly different. The actual length of the WPM symbols is defined by the wavelet used and in general it is significantly longer than the OFDM symbol. In the case of time offset this overlap of the symbols in WPM causes each symbol to interfere with several other symbols, while in OFDM the symbol only interferes with the one adjacent symbol.

The second important difference between the two transmission schemes is the use of the guard interval between the symbols. OFDM uses cyclic prefix that significantly improves its performance under loss of time synchronization[3]. The WPM, on the other hand, cannot benefit from such guard intervals since the WPM symbols overlap one another.

[3] However, the use of cyclic prefix is effective only when the time offset induced by the channel does not exceed the length of the cyclic prefix and that the direction of time shift is towards the cyclic prefix.

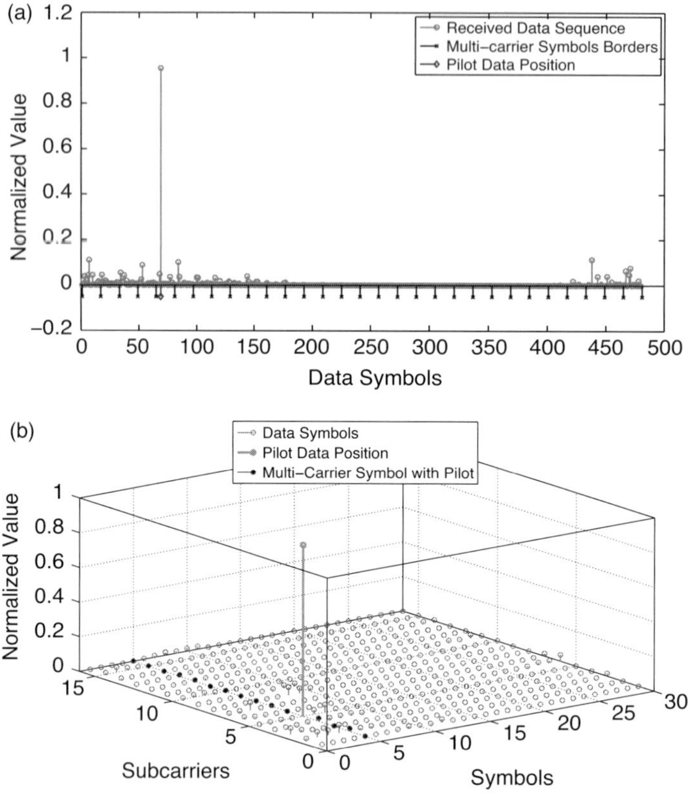

Figure 4.20 Spectral energy in a frame of received WPM (with Daubechies wavelet) signal affected by phase noise. (a) 2D and (b) 3D view.

4.4.2 Time offset in OFDM

Cyclic prefix is an effective and low complexity method to cope with dispersive channels and time synchronization errors in OFDM transceivers. OFDM is often accommodated with cyclic prefix but rarely with cyclic postfix. This means that we have two distinct situations that can occur under time synchronization errors, depending on the direction of the time offset [18].

- time synchronization error away from cyclic prefix (to the right);
- time synchronization error towards cyclic prefix (to the left).

4.4.2.1 Time offset away from the cyclic prefix

Figure 4.21 illustrates this example by considering a snapshot of OFDM data transmission with three OFDM symbols ($u-1, u, u+1$). In this example the FFT window (for data demodulation) is misaligned to the right, i.e., away from the cyclic prefix. Each OFDM symbol consists of N data samples and an extension of N_{CP} samples representing cyclic prefix. The FFT window, in the case considered, will contain $N - \Delta_t$ data samples $((\Delta_t + 1), (\Delta_t + 2), \ldots N)$ of the uth OFDM symbol, missing the first Δ_t useful

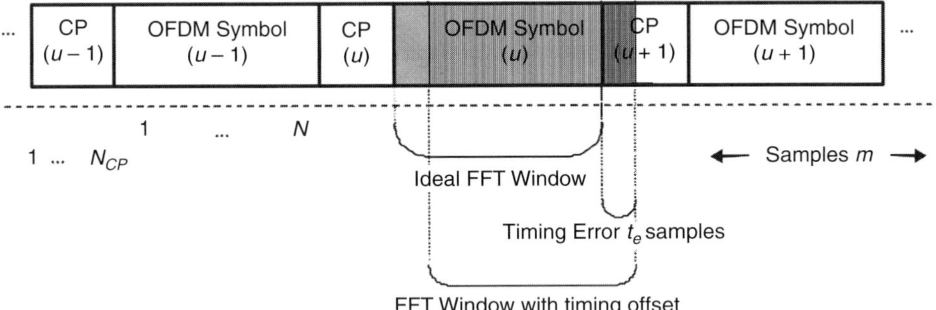

Figure 4.21 Timing offset away from cyclic prefix (to the right).

samples. Instead Δ_t samples $(1, 2, \ldots \Delta_t)$ of the next $(u+1)$th OFDM symbol will be erroneously selected.

An OFDM system that is affected by timing errors and where samples of neighbouring symbol are wrongly selected experiences severe degradation of the performance. The demodulated OFDM signal after FFT can be written as:

$$\hat{a}_{u',k'} = \underbrace{\frac{N-\Delta_t}{N} a_{u',k'} e^{j2\pi\frac{k'}{N}\Delta_t}}_{\text{Useful Data (Attenuated, phase shifted)}} + \underbrace{\frac{1}{N}\sum_{n=0}^{N-1-\Delta_t}\sum_{k=0;k\neq k'}^{N-1} a_{u',k} e^{j2\pi\frac{k(n+\Delta_t)}{N}} e^{-j2\pi\frac{k'n}{N}}}_{\text{Inter-Carrier Interference}}$$
$$+ \underbrace{\frac{1}{N}\sum_{n=N-\Delta_t}^{N-1}\sum_{k=0}^{N-1} a_{u+1,k} e^{j2\pi\frac{k(n-N+\Delta_t)}{N}} e^{-j2\pi\frac{k'n}{N}}}_{\text{Inter-Symbol Interference}} \tag{4.29}$$

The first component of Eq. (4.29) represents useful signal that is attenuated and phase shifted by a term proportional to sub-carrier index k'. The second component of Eq. (4.29) gives ICI and the third component stands for ISI with the next symbol.

4.4.2.2 Time offset towards the cyclic prefix

The other situation occurs when we have time offset towards the symbol's own cyclic prefix, i.e., to the left. Figure 4.22 illustrates such a scenario. In this case FFT window consists of first $N - \Delta_t$ samples $(1, 2, \ldots (N - \Delta_t))$ of the considered uth OFDM symbol and the last Δ_t samples of the own cyclic prefix. We assume for the convenience that $\Delta_t < N_{CP}$.

The demodulated OFDM signal affected by time offset in the direction of symbol's own cyclic prefix is given in Eq. (4.30), for the case when $\Delta_t < N_{CP}$.

$$\hat{a}_{u',k'} = a_{u',k'} e^{-j2\pi\frac{k'\Delta_t}{N}} \tag{4.30}$$

Thanks to the cyclic prefix, the orthogonality is preserved and ISI and ICI terms do not appear. The timing error towards the cyclic prefix results therefore in pure phase shift.

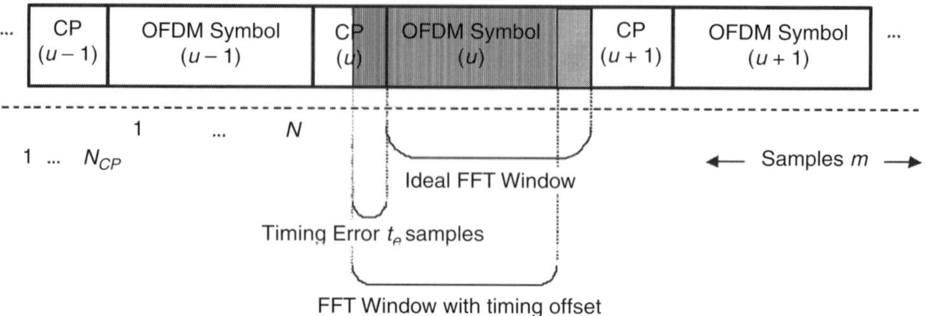

Figure 4.22 Timing offset towards cyclic prefix (to the left).

4.4.3 Time synchronization error in WPM

The WPM transceivers do not employ guard intervals and therefore the direction of time offset is inconsequential. The demodulation process under time offset error Δ_t can be derived as:

$$\begin{aligned}\hat{a}_{u',k'} &= \sum_n R[n]\xi_l^{k'}[(u'N - n + \Delta_t] = \sum_n \sum_u \sum_{k=0}^{N-1} a_{u,k}\xi_l^k[n - uN]\xi_l^{k'}[u'N - n + \Delta_t] \\ &= \sum_u \sum_{k=0}^{N-1} a_{u,k}\left(\sum_n \xi_l^k[n - uN]\xi_l^{k'}[u'N - n + \Delta_t]\right)\end{aligned} \tag{4.31}$$

For ease of representation we define the cross-waveform function $\Omega[n]$ as:

$$\Omega_{k,k'}^{u,u'}[\Delta_t] = \sum_n \xi_l^k[n - uN]\xi_l^{k'}[u'N - n + \Delta_t] \tag{4.32}$$

Expression (4.32) represents the autocorrelation and the cross-correlation of the WPM waveforms, depending on sub-carrier index k. When $k = k'$, the two sub-carrier waveforms are time-inversed versions of one another and Eq. (4.32) gives the autocorrelation sequence of the waveform k. In the other cases when $k \neq k'$, the two waveforms correspond to different sub-carriers and Eq. (4.32) stands for the cross-correlation between waveforms k and k'.

Using Eq. (4.31) and Eq. (4.32), we can now express the demodulated alphabet for the kth sub-carrier and uth WPM symbol corrupted by the interference due to loss of orthogonality as:

$$\hat{a}_{u',k'} = \underbrace{a_{u',k'}\Omega_{k'k'}^{u',u'}[\Delta_t]}_{\text{Desired Alphabet}} + \underbrace{\sum_{u;u\neq u'} a_{u,k'}\Omega_{k',k'}^{u,u'}[\Delta_t]}_{\text{ISI}} + \underbrace{\sum_u \sum_{k=0;k\neq k'}^{N-1} a_{u,k}\Omega_{k,k'}^{u,u'}[\Delta_t]}_{\text{IS-ICI}} \tag{4.33}$$

In Eq. (4.33) the first term stands for attenuated useful signal. The second term gives the ISI due to symbols transmitted on the same sub-channel and the third term denotes ICI measured over the whole frame.

The received constellation points of WPM under time synchronization errors do not experience linear phase rotation, as opposed to OFDM where rotation of constellation points is proportional to the sub-carrier index. The WPM signal in the presence of timing error will, however, be attenuated and it will suffer from ISI and ICI.

4.4.4 Modulation scheme

The consequence of time-offset in OFDM, regardless of offset direction, is the introduction of phase shift [18]. The phase shift is linearly proportional to the sub-carrier index and the value of the time offset. The rotation angle $\Phi_t[k]$ due to a timing error is given by [18]:

$$\Phi_t[k] = \frac{2\pi k \Delta_t}{N} \tag{4.34}$$

Standard modulation techniques such as coherent (non-differential) quadrature phase shift keying (QPSK) perform poorly under time synchronization error because the sub-carriers with higher-frequency indices experience greater phase shifts. And even a small timing offset (such as $\Delta_t = 1$) results in a phase rotation of constellation symbols of the order of $0 \leq \Phi_t[k] < 2\pi$. The sub-carrier with the highest frequency will therefore experience a phase shift of almost 360 degrees. If this phase shift is not corrected, majority of the detected data would be corrupted even without ICI or ISI.

The phase rotation due to timing errors can be usually revised by pilot-symbol-aided channel estimation techniques or by use of differential constellation mapping. In this work we employ differential quadrature phase shift keying (DQPSK) in order to overcome this problem. In the DQPSK scheme the data are modulated on the basis of the phase difference between two consecutive constellation symbols, thereby ensuring that adjacent sub-carriers experience a phase shift that is independent of the carrier position. The phase rotation of constellation point k is determined by applying a phase shift of $\Delta\Phi$ to the previous constellation symbol $k - 1$. The difference in phase shift $\Delta\Phi$ is determined by the unmodulated data value assigned to sub-carrier k, so in the case of DQPSK the phase shift can be written as:

$$\Delta\Phi_b = \frac{2(b-1)\pi}{4}, \qquad b \in 1, \ldots, 4 \tag{4.35}$$

The phase difference between two consecutive DQPSK constellation symbols under timing errors becomes:

$$\Delta\phi_{k,k-1} = e^{j\left(\Delta\Phi_b - \frac{2\pi\Delta_t}{N}\right)} \tag{4.36}$$

Using DQPSK modulation in the presence of a timing error therefore results in a phase shift that is dependent on the value of the time offset but no longer on the value of the sub-carrier index k. The rotation angle $\Phi_t[k]$ due to the timing error becomes:

$$\Phi_t[k] = \frac{2\pi\Delta_t}{N}, \quad \text{differential PSK} \tag{4.37}$$

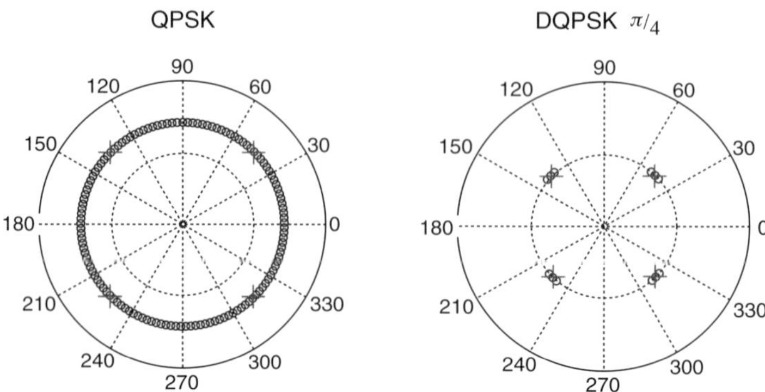

Figure 4.23 Constellation plots of received OFDM signal with a timing error of $\Delta_t = 1$; Left: QPSK, Right: DQPSK $^{\pi}/_{4}$.

Figure 4.23 illustrates the rotations of constellation points for a received OFDM signal using QPSK and DQPSK modulation. We have assumed here an ideal channel and a time offset of $\Delta_t = 1$ samples towards the cyclic prefix.

DQPSK modulation is a simple solution to overcome the problem of phase shift under time synchronization errors. However, DQPSK modulation requires about 2 to 3 dB higher SNR when compared to coherent QPSK to obtain the same BER performance in an AWGN channel.

4.4.5 Numerical results for time offset

The performances of WPM and OFDM in the presence of timing synchronization errors are investigated by means of computer simulations. The time offset is modelled as a discrete uniform distribution between –2 and 2 samples, i.e., $\Delta_t \in [-2, -1, 0, 1, 2]$.

The designed parameters deviate to some extent from those used for frequency offset and phase noise. First, the DQPSK modulation scheme is used instead of QPSK modulation in order to prevent OFDM constellation points experiencing large phase shifts. Secondly, in OFDM we use cyclic prefix of 16 samples that is placed in front of OFDM symbols; in WPM we do not use any guard interval. Due to utilization of cyclic prefix the spectral efficiency of OFDM is decreased by 12.5%, while the spectral efficiency of WPM has remained unchanged. Finally, we oversample the data to magnify the difference in performance between various systems and wavelets.

An overview of simulation parameters is given in Table 4.3.

4.4.5.1 Performance of WPM without time errors

To gauge the system operation we first check its performance under optimal conditions. Figure 4.24 shows the DQPSK constellation points for WPM setup using various wavelets in ideal channel conditions with no time offset. From the plot we may note that perfect estimates of the transmitted data can be obtained at the receiver when the transmitter and receiver ends are in time unison.

Table 4.3 Simulation setup for evaluation of performance under time synchronization error

	WPM	OFDM
Number of sub-carriers	128	128
Number of multi-carrier symbols per frame	100	100
Modulation	DQPSK	DQPSK
Channel	AWGN	AWGN
Oversampling factor	15	15
Guard band	–	–
Guard interval	–	CP (length: 16)
Frequency offset	–	–
Phase noise	–	–
Time offset	$\Delta_t \in [-2, -1, 0, 1, 2]$ (Uniformly Distributed)	$\Delta_t \in [-2, -1, 0, 1, 2]$ (Uniformly Distributed)

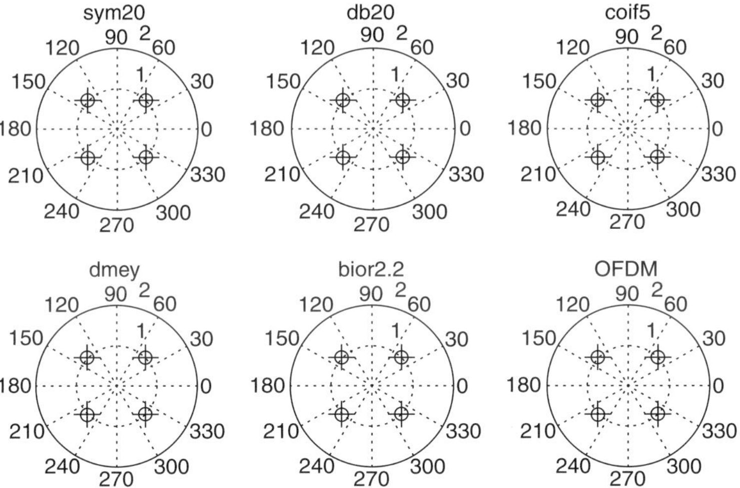

Figure 4.24 DQPSK constellation points for WPM setup using various wavelets in ideal channel with no time offset.

A timing error results in a loss of time synchronization that causes a loss of orientation of incoming data at the receiver. As a result, the time-domain data entering the IDFT/IDWT block is incorrectly aligned whereby the samples of previous or next OFDM/WPM symbol are selected, while valid samples at the beginning or at the end of the symbol in consideration are discarded. We present the impact of time synchronization error in the next sections.

4.4.5.2 Performance when time offset is modelled as discrete uniform distribution

Figure 4.25 illustrates the bit error rates (BER) of OFDM and WPM transceivers over AWGN channel for uniformly distributed timing offset of $\Delta_t = 2$ samples. The OFDM

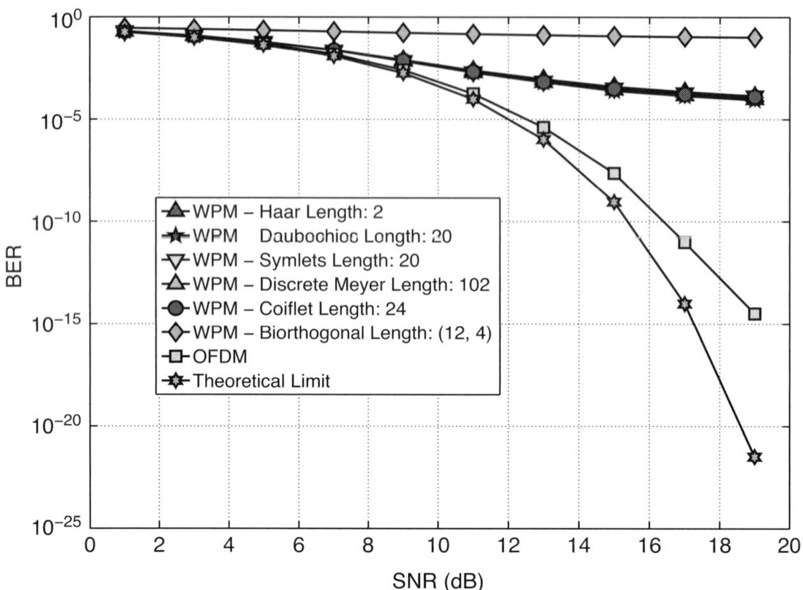

Figure 4.25 BER for WPM with different wavelets and OFDM under time synchronization errors ($\Delta_t = 2$).

system performs much better under time synchronization errors when compared to the WPM, partly due to exploiting cyclic prefix and cancellation of phase rotation by the DQPSK modulation scheme. The WPM cannot profit from these revisions and therefore show poor performance in the presence of timing error. Once again biorthogonal wavelet has the highest BER due to an unfulfiled perfect reconstruction condition.

In Figure 4.26 the BER is shown for different values of time offset varying from –15 to 12 samples. The time offset in this simulation is modelled as a one-sided uniform distribution varying in order to highlight the importance of cyclic prefix for OFDM. The value of time offset is considered to take values $\{0, \ldots, \Delta_t\}$ for timing errors to the right and $\{-\Delta_t, \ldots, 0\}$ for the timing errors to the left. The SNR is kept constant at 10 dB.

The direction of time offset is inconsequential for WPM systems, as can be seen at Figure 4.26. The BER curves of WPM are almost perfect mirror images with respect to the origin. This does not hold for OFDM, since we can see clearly that the negative timing offset (towards the own cyclic prefix) results in much lower BER when compared to the positive timing offsets (away from the own cyclic prefix). Due to the use of cyclic prefix the misalignment of FFT window between the boundaries of extended symbols does not cause interference. However, when time offset exceeds the length of cyclic prefix the ICI and ISI terms reappear.

4.4.5.3 Influence of number of sub-carriers

Figure 4.27 shows the performance of the WPM in the presence of time synchronization error when the number of sub-carriers is altered. The plots reveal how the performance of

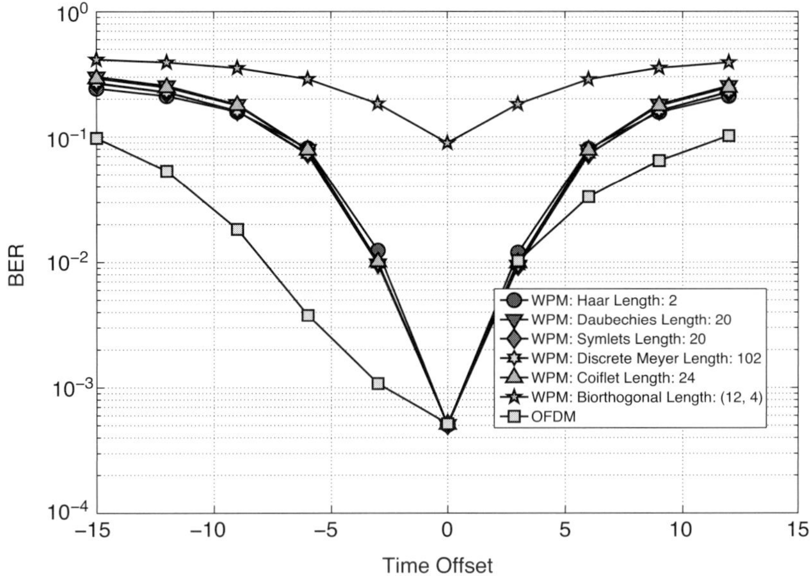

Figure 4.26 BER vs. time offset error for WPM and OFDM in AWGN channel (SNR = 10 dB).

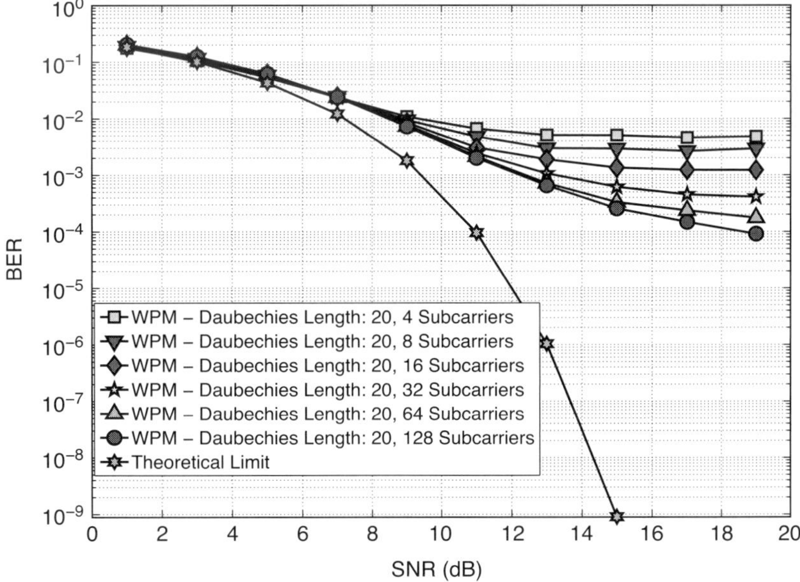

Figure 4.27 BER plots for WPM transmission under a loss of time synchronization for different number of sub-carriers.

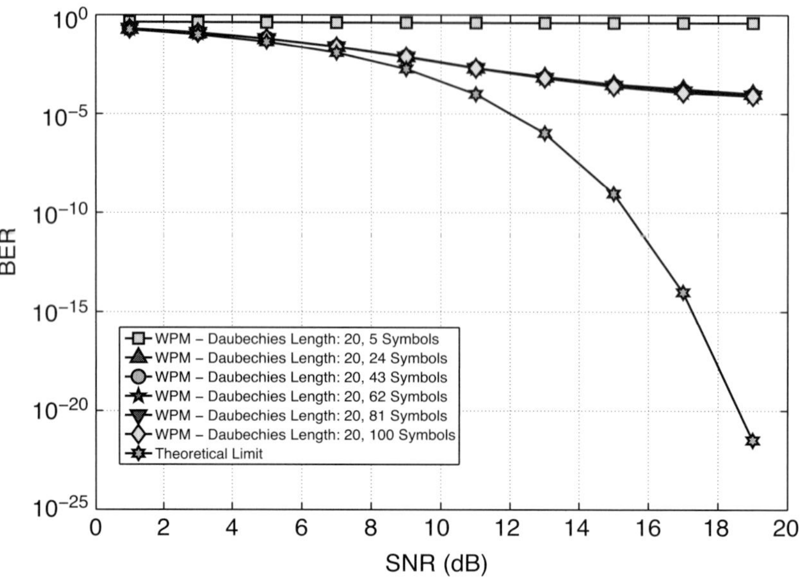

Figure 4.28 BER for WPM with timing error for different number of symbols/frame.

WPM under a time offset error depends on the number of sub-carriers. With increasing number of WPM sub-carriers the BER significantly decreases. We recall from Chapter 3 that the symbol duration of multi-carrier system is proportional to the number of sub-carriers used. Therefore, a larger number of sub-carriers means longer symbol duration and hence smaller relative time offset with respect to multi-carrier symbol length.

4.4.5.4 Influence of number of symbols/frame

The simulation results for different numbers of WPM symbols per frame are depicted in Figure 4.28. The results show that the number of symbols/frame does not influence the BER performance.

4.4.5.5 Influence of different lengths of wavelet filters

Figure 4.29 illustrates the influence of filter length and number of zero wavelet moments in combination with timing error on the BER. The wavelet family of choice is Daubechies. In the plots, the Daubechies filter with 6 coefficients and 3 wavelet zero moments appears to have a slightly higher BER when compared to longer filters from the same family. However, when the length of the filters is further increased the BER curves become closely spaced, and therefore we can conclude that there is no significant relation between the BER and filter length in the presence of timing errors.

4.4.5.6 Influence of number of sub-carriers

The effect of time synchronization error on the constellation points is depicted in Figure 4.30. In order to highlight the effect of time synchronization error, we

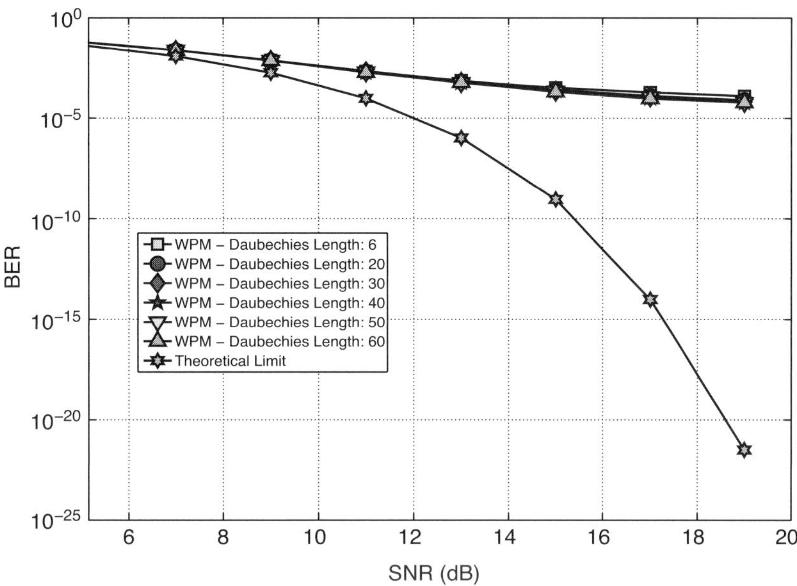

Figure 4.29 BER for WPM using Daubechies wavelets of different lengths under a loss of time synchronization.

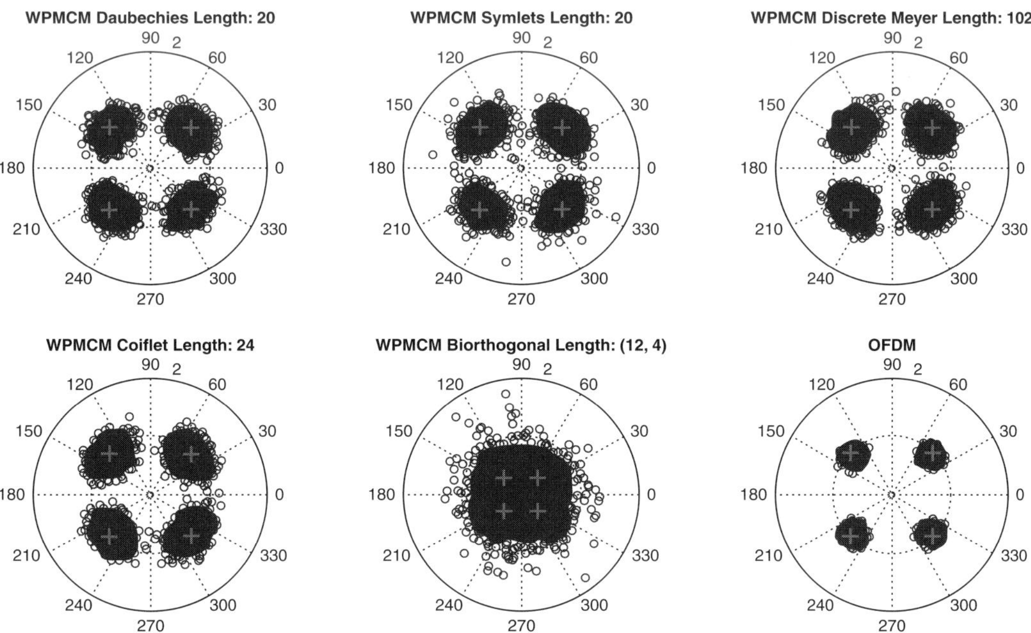

Figure 4.30 Constellation points of received signal in the presence of timing error.

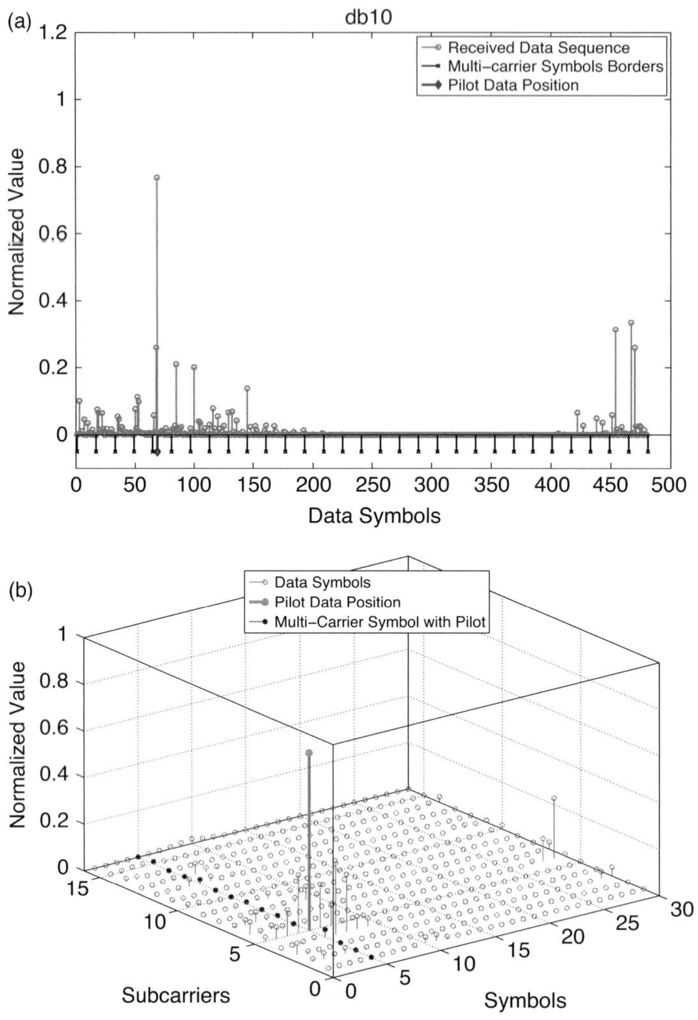

Figure 4.31 Spectral energy in a frame of received WPM (with Daubechies wavelet) signal affected by timing errors. (a) 2D, (b) 3D view.

consider an ideal channel without any noise (apart from a time offset error). The main impact of the time offset is the scattering of the constellation points around reference positions due to interference. OFDM has more concentrated constellation points than any tested WPM system, which indicates that the signal to interference ratio (SIR) of OFDM is higher than SIR of WPM under timing errors.

4.4.5.7 Influence of number of sub-carriers

Figures 4.31 and 4.32 illustrate the spreading of sub-carrier energy due to time synchronization error for WPM and OFDM systems, respectively. The timing error in OFDM

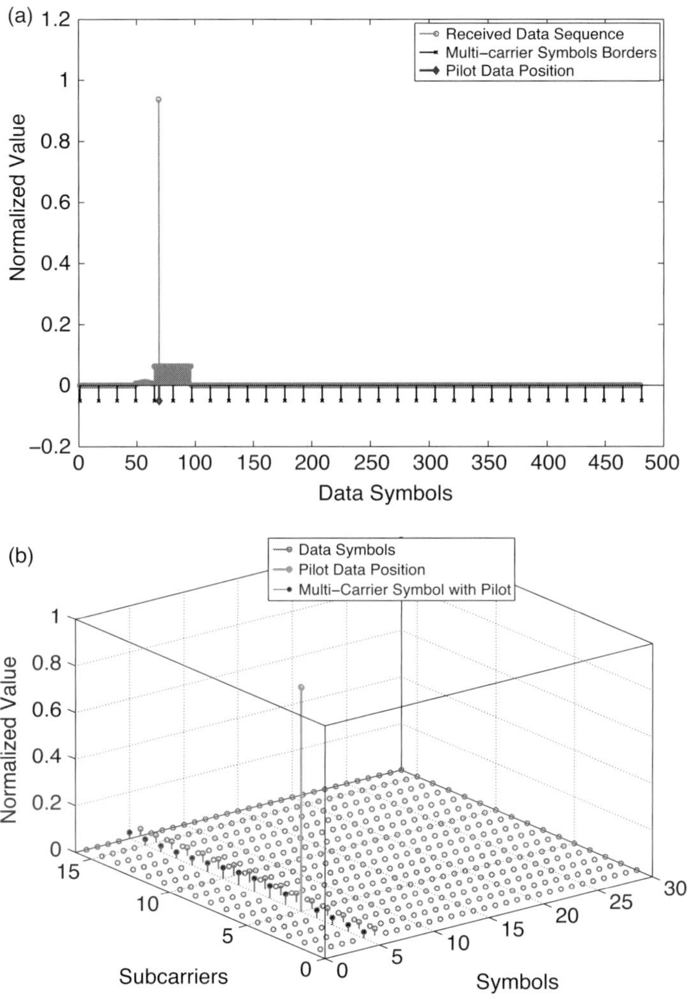

Figure 4.32 Spectral energy in a frame of received OFDM signal affected by timing errors. (a) 2D, (b) 3D view.

results in ISI between successive symbols and ICI. If the cyclic prefix is used, the ICI and ISI terms are cancelled for time offset towards the symbol's own cyclic prefix, i.e., time offset to the left. This can also be seen in Figure 4.32 where the energy of the pilot symbol is spread into the subsequent symbol (ISI) but not into the previous symbol. Furthermore, the energy of a pilot sub-carrier is also spread across the other sub-carriers located in the same symbol (ICI).

In the case of WPM the energy of a single pilot is spread across a number of symbols, where sub-carriers close to the pilot sub-carrier contain the greatest part of interfering energy, regardless of the symbol index.

4.5 Summary

In this chapter we addressed the effects of frequency offset, phase noise and time synchronization error on WPM and OFDM transceivers. The frequency offset and phase noise lead to the loss of orthogonality and sub-carriers begin to interfere with each other. In OFDM the disturbance is limited to generation of ICI, but in WPM the frequency offset and phase noise cause ICI as well as inter-symbol ICI. The effect of time synchronization error was also discussed. Akin to OFDM, we found that the timing error in WPM contains two components: ISI and the inter-symbol ICI. There are, however, significant differences between both schemes in presence of timing errors. First, the ISI in OFDM arises only between successive symbols while in WPM a number of symbols interfere one with another. Secondly, the timing error in OFDM results in a rotation of constellation symbols that is proportional to the sub-carrier frequency, but in WPM this behaviour is absent.

The effects of frequency offset, phase noise and time synchronization errors were also examined by the simulation studies. Several well-known wavelets such as Daubechies, Symlets, discrete Meyer, Coiflet and biorthogonal wavelet were applied and studied. The sensitivities of WPM and OFDM are quite similar in the presence of frequency offset and phase noise. However, time synchronization error is found to be a major drawback of a WPM transceiver. The simulations have shown that OFDM has much lower BER under timing errors when compared to WPM. This is largely due to the beneficial use of cyclic prefix in OFDM.

The wavelets used in these computer simulations are standard wavelets that were developed for other applications such as image processing or compression, and hence are not suitable for telecommunication purposes. In Chapter 7 we will present the design of new wavelets that minimize the interference due to time offset errors [30]–[32].

References

[1] G. Wornell, "Emerging Applications of Multirate Signal Processing and Wavelets in Digital Communications," *Proceedings of IEEE*, vol. 84, pp. 586–603, April 1996.

[2] M.K. Lakshmananm and H. Nikookar, "A Review of Wavelets for Digital Wireless Communication," *Springer Journal on Wireless Personal Communication*, vol. 37, No. 3–4, May 2006, pp. 387–420.

[3] A. Lindsey, "Wavelet Packet Modulation for Orthogonally Transmultiplexed Communications," *IEEE Transaction on Signal Processing*, vol. 45, May 1997, pp. 1336–9.

[4] T. Pollet, M. van Bladel, and M. Moeneclaey, "BER Sensitivity of OFDM Systems to Carrier Frequency Offset and Wiener Phase Noise," *IEEE Transaction on Communications*, vol. 43, pp. 191–3, April 1995.

[5] A.G. Armada, "Understanding the Effect of Phase Noise in OFDM," *IEEE Transactions on Broadcasting*, vol. 47, No. 2, pp. 153–9, June 2001.

[6] H. Nikookar and R. Prasad, "On the Sensitivity of Multi-carrier Transmission over Multipath Channels to Phase Noise and Frequency Offset," in *Proc. 7th IEEE International Symposium*

Personal, Indoor Mobile Radio Communication (PIMRC '96), vol. 1, pp. 68–72, October 1996.

[7] T.C.W. Schenk, R.W. van der Hofstad, and E.R. Fledderus, "Distribution of the ICI Term in Phase Noise Impaired OFDM Systems," *IEEE Transactions on Wireless Communications*, vol. 6, No. 4, pp. 1488–500, April 2007.

[8] H. Steendam and M. Moeneclaeym, "Sensitivity of Orthogonal Frequency Division Multiplexed Systems to Carrier and Clock Synchronization Errors," *Elsevier Signal Processing*, pp. 1217–29, November 1997.

[9] L. Tomba, "On the Effect of Wiener Phase Noise in OFDM Systems," *IEEE Transaction on Communications*, vol. 46, No. 5, May 1998.

[10] K. Sathananthan and C. Tellambura, "Probability of Error Calculation of OFDM System with Frequency Offset," *IEEE Transactions on Communications*, vol. 49, No. 11, pp. 1884–8, November 2001.

[11] H. Nikookar and B.G. Negash, "Frequency Offset Sensitivity Reduction of Multi-carrier Transmission by Waveshaping," in *Proc. IEEE International Conference on Personal Wireless Communications (ICPWC '2000)*, pp. 444–8, December 2000.

[12] J. Armstrong, "Analysis of New and Existing Methods of Reducing Intercarrier Interference Due to Frequency Offset in OFDM," *IEEE Transaction on Communications*, vol. 47, No. 3, pp. 365–9, March 1999.

[13] P.H. Moose, "A Technique for Orthogonal Frequency Division Multiplexing Frequency Offset Correction," *IEEE Transaction on Communications*, vol. 42, No. 10, pp. 2908–14, October 1994.

[14] A.G. Armada and M. Calvo, "Phase Noise and Sub-Carrier Spacing Effects on the Performance of an OFDM Communication System," *IEEE Communication Letters*, vol. 2, No. 1, pp. 11–13, January 1998.

[15] Y.C. Lion and K.C. Chen, "Estimation of Wiener Phase Noise by the Autocorrelation of the ICI Weighting Function in OFDM System," in *Proc. 16th IEEE International Symposium on Personal, Indoor and Mobile Radio Communications (PIMRC '05)*, pp. 725–9, 2005.

[16] P. Robertson and S. Kaiser, "Analysis of the Effects of the Phase Noise in Orthogonal-Frequency Division Multiplexed (OFDM) Systems," in *Proc. IEEE International Conference on Communications (ICC '95)*, vol. 3, pp. 1652–7, June 1995.

[17] V.S. Abhayawardhana and I.J. Wassell, 'Common Phase Error Correction with Feedback for OFDM in Wireless Communication," in *Proc. IEEE Globecom*, vol. 1, pp. 651–5, November 2002.

[18] C.R.N. Athaudage, "BER Sensitivity of OFDM Systems to Time Synchronization Error," in *Proc. IEEE International Conference on Communication Systems (ICCS 2002)*, vol. 1, pp. 42–6, November 2002.

[19] Y. Mostofi and D.C. Cox, "Mathematical Analysis of the Impact of Timing Synchronization Errors on the Performance of an OFDM System," *IEEE Transactions on Communications*, vol. 54, No. 2, February 2002.

[20] D. Liu, Y. Tang, D. Sang, and S. Li, "Impact of the Timing Error on BER Performance of TDD Pre-equalized OFDM System," in *Proc. IEEE International Conference on Personal Wireless Communications (ICPWC '2000)*, vol. 1, pp. 714–18, September 2004.

[21] D. Lee and K. Cheun, "Coarse Symbol Synchronization Algorithms for OFDM Systems in Multipath Channels," *IEEE Communication Letters*, vol. 6, No. 10, pp. 446–8, October 2002.

[22] M. Sandell, J.J. Beekm, and P.O. Brjesson, "Timing and Frequency Synchronization in OFDM Systems Using the Cyclic Prefix," *International Symposium on Synchronization*, pp. 16–19, 1995.

[23] M. Tanda, "Blind Symbol-timing and Frequency-offset Estimation in OFDM Systems with Real Data Symbols," *IEEE Transactions on Communications*, vol. 52, No. 10, pp. 1609–12, October 2004.

[24] A.J. Coulson, "Maximum Likelihood Synchronization for OFDM using a Pilot Symbol: algorithms," *IEEE Journal on Selected Areas in Communications*, vol. 19, No. 12, pp. 2486–94, December 2001.

[25] B. Yang, K.B. Letaief, R.S. Cheng, and Z. Cao, "Timing Recovery for OFDM Transmission," *IEEE Journal on Selected Areas in Communications*, vol. 18, No. 11, pp. 2278–91, November 2000.

[26] A.B. Narasimhamurthy, M.K. Banavar, and C. Tepedelenliouglu, *OFDM Systems for Wireless Communications*, Morgan and Claypool Publishers, 2010.

[27] T. Schenk, *RF Imperfections in High-rate Wireless Systems*, Springer, 2008.

[28] A. Demir, A. Mehrotra, and J. Roychowdhury, "Phase Noise in Oscillators: a Unifying Theory and Numerical Methods for Characterization," *IEEE Transactions on Circuits and Systems*, pp. 655–74, May 2000.

[29] R. Corvaja and S. Pupolin, "Phase Noise Limits in OFDM Systems," in *Proc. International Wireless Personal Multimedia Communications Symposium (WPMC '03)*, pp. 19–22, October 2003.

[30] M.K. Lakshmanan, D. Karamehmedovic, and H. Nikookar, "An Investigation on the Sensitivity of Wavelet Packet Modulation to Time Synchronization Error," *Special Issue Wireless Personal Communications*, Springer, 2010.

[31] D. Karamehmedović, M.K. Lakshmanan, and H. Nikookar, "Performance Evaluation of WPMCM with Carrier Frequency Offset and Phase Noise," *Journal of Communications (JCM)*, vol. 4, no. 7, Aug. 2009, pp. 496–508.

[32] D. Karamehmedović, M.K. Lakshmanan, and H. Nikookar, "Optimal Wavelet Design for Multicarrier Modulation with Time Synchronization Error," *IEEE GlobeCom 2009*, Hawaii, Dec. 2009.

5 Peak-to-average power ratio

5.1 Background

Wavelet packet modulation (WPM) is a new transmission technique that is seen as a promising alternative to the well-established orthogonal frequency division multiplexing (OFDM). The main advantage of WPM is that the transmission characteristics of the system can be adapted according to the radio environment to maximize resource utilization. However, as in OFDM, WPM transmission also suffers from large power fluctuations and high peak-to-average power ratio (PAPR). Multi-carrier systems such as WPM or the classical OFDM combine many independently modulated sub-carriers to obtain a composite signal. If sub-carriers add coherently, then the peak power of the composite OFDM/WPM signal can be many times larger than the average power. This can lead to non-linear distortions and degradation of system performance. It is therefore important to study the power fluctuations associated with multi-carrier systems and mitigate it. In this chapter we address the PAPR performance of the WPM system. We first study the stochastic nature of the WPM signal, its power variations and its PAPR performance. We then study suitable strategies that mitigate PAPR.

5.2 Introduction

OFDM and WPM belong to a class of transmission systems known as the multi-carrier systems. The signals in such systems are modulated using different sub-carrier frequencies and transmitted simultaneously. An inherent disadvantage of such multi-carrier systems is the high peak-to-average power ratio (PAPR) of the transmitted signals. A multi-carrier modulation (MCM) signal consists of a number of independently modulated sub-carriers, which can give a large PAPR when added up coherently. In fact, the peak of the signal can be up to M times the average power (where M is the number of sub-carriers) if all the sub-carriers are of the same phase.

Since practical systems are limited by the maximum operable power, either the WPM/OFDM systems have to work with a large power back-off or operate in the non-linear (saturation) region of the electronic components in the transceiver chain such as the power amplifier and digital-to-analogue converters. A large back-off would mean that the average signal power must be kept low to ensure that the amplifier operates in the linear region. However, this will have a detrimental effect on the efficiency of power utilization, particularly in mobile systems where battery lifetime is a premium resource. On the other hand, operating in the nonlinear region of the amplifiers can result in intermodulation distortions and an ensuant increase in error rates. It is therefore important to minimize the power fluctuations.

The WPM is a developmental system and has not been studied extensively, especially its PAPR performance. While the quantity of literature available for the study of OFDM and its PAPR performance is significant, the material available for a similar study on WPM is scarce. In fact, the entire material existing on the subject can be listed as follows: In [1] a study on the PAPR of WPM signals and its stochastic variations is presented. The study shows that the envelope of the WPM signal is Gaussian and its power is chi-square distributed. Furthermore, the PAPR performances of the WPM systems for almost all used wavelets are shown to be similar to OFDM. In [2] a multi-pass pruning method to reduce PAPR is proposed. And in [3] a threshold-based method to reduce PAPR is suggested. In [4] upper bounds for the maximum PAPR for WPM transmission are derived and based on these results wavelets that minimize PAPR are obtained. A different approach is followed in [5] where the tree structure, used to derive the sub-carriers, is adjusted to lower the PAPR.

In this chapter, we present our study on the PAPR performance of WPM systems where the stochastic of the WPM signals and their power is investigated. Furthermore, the variation of PAPR with different wavelets and pulse shapes is gauged. Subsequently, the influence of the pulse-shaping characteristics on the PAPR is investigated. This study shows that the envelope of the WPM signal is Gaussian and its power distribution is chi-squared. Several well-known wavelets such as Daubechies, Symlets, Coiflets, discrete Meyer and biorthogonal wavelet are considered and the PAPR performances are compared. The chapter is organized as follows. In Section 5.3 a study on the nature of WPM signals is presented. A brief survey on existing PAPR reduction techniques is provided in Section 5.4. In Section 5.5 PAPR reduction by modification of sub-carrier phases is presented and finally the contents of the chapter are summarized in Section 5.6.

5.3 PAPR distribution of multi-carrier signal

5.3.1 OFDM

A multi-carrier signal consists of a number of independently modulated sub-carriers, which can give a large peak-to-average-power ratio when they add up coherently. In fact, for a system with M sub-carriers, with all of them of the same phase, the peak power of the transmitted signal can be M times the average power. Let X_m, $m = 0, 1, \ldots,$ $M - 1$ be the information bits expressed as a vector $\mathrm{X} = [X_0, X_1, \ldots X_{M-1}]$, then the continuous time-varying representation of a single complex baseband OFDM symbol with M sub-carriers, can be written as:

$$x(t) = \frac{1}{\sqrt{M}} \sum_{m=0}^{M-1} X_m e^{j2\pi m\Delta f t}, \quad 0 \leq t \leq T \tag{5.1}$$

where $j = \sqrt{-1}$, Δf is the sub-carrier spacing and T is the symbol duration. The orthogonality is established with a sub-carrier spacing $\Delta f = 1/T$. And the discrete form of the baseband signal can be written as:

$$x[n] = \frac{1}{\sqrt{M}} \sum_{m=0}^{M-1} X_m e^{\frac{j2\pi m}{M}}, \quad n = 0, 1, \ldots, M-1 \tag{5.2}$$

The PAPR is one way to measure the variation of the transmitted signal about its mean and for critically sampled data $x[n]$ can be given as:

$$\text{PAPR} = \frac{\max_{0 \le n \le M-1}(|x[n]|^2)}{E(|x[n]|^2)} \tag{5.3}$$

where $E(|.|)$, the expected value, denotes the average.

The cumulative distribution function (CDF) of the PAPR is one of the most frequently used performance measures for PAPR reduction techniques, [6], [7]. From the central limit theorem it follows that for large number of sub-carriers M, the real and imaginary components of $x[n]$ follow the Gaussian distribution, each with a zero mean and variance of M times the variance of one complex sinusoid. The amplitude of the OFDM signal therefore has a Rayleigh distribution and its power distribution becomes a central chi-square distribution with two degrees of freedom and zero mean [6], [8]. The CDF of the power is given as [6]

$$F(z) = \int_0^z \frac{1}{2\sigma^2} e^{-\frac{u}{2\sigma^2}}\, du = 1 - e^{-\frac{z}{2\sigma^2}} \tag{5.4}$$

where $z \ge 0$. From the power distribution the theoretical CDF for PAPR per OFDM symbol can be derived. Assuming the samples to be mutually uncorrelated (which is true when there is no oversampling), the probability that PAPR is below some threshold level z, can be written as

$$Prob\{\text{PAPR} \le z\} = [F(z)]^M = \left(1 - e^{-\frac{z}{2\sigma^2}}\right)^M \tag{5.5}$$

5.3.2 WPM

A WPM signal, like the OFDM signal, is the sum of many information bearing sub-carriers which are statistically independent. The orthogonal sub-carriers are wavelet packet bases derived from multi-resolution analysis (MRA) [9] as explained in Chapter 3.

5.3.2.1 WPM signal characteristics

In Figure 5.1 the CDF curves for the PAPR of the WPM system (theoretical as well as simulated values) for different numbers of sub-carriers M are plotted. We can see from Figure 5.1 that as the number of carriers of the WPM system is increased, the simulated and the theoretical (as represented in Eq. (5.5)) PAPR curves converge. From about $M > 128$ carriers the theoretical derivations accurately map the simulated values.

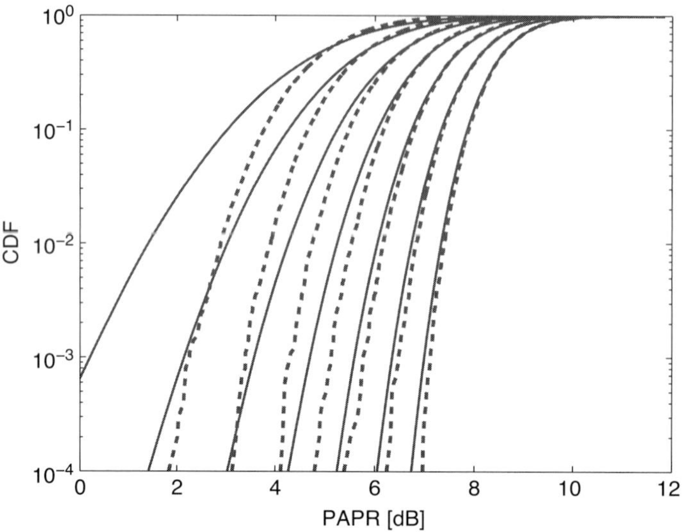

Figure 5.1 CDF distribution of PAPR for different number of sub-carriers. From left to right: 16, 32, 64, 128, 256, 512, 1024 (dashed lines indicate simulated values while continuous lines present the theoretical curves). Wavelet of choice is Daubechies 15.

In the following set of simulations we consider a WPM system realized using a filter bank structure with 7 levels of decomposition yielding 128 carriers.

5.3.2.2 Amplitude distribution

Unlike OFDM, which is a complex signal with real and imaginary parts, the WPM signal only has real components. An OFDM signal has a Rayleigh distribution and it would be interesting to check the distribution of WPM signal. Figure 5.2 plots the simulated CDF curves for WPM systems along with Gaussian and Rayleigh distributions. The WPM setup uses Daubechies wavelets with length 15. It is clear from the figure that the patterns of the WPM signal variations follow the Gaussian distribution.

5.3.2.3 Power distribution

The central limit theorem (CLT) states that the mean of a sufficiently large number of independent random variables, each with finite mean and variance, will be normally distributed. Based on CLT, when a large number of sub-carriers are employed in a WPM system; i.e., a large number of levels in the IDWPT, the amplitude of the WPM signal follows a Gaussian distribution. It is well known from the stochastic theory that the distribution of power of Gaussian signals is chi-squared. This means that the power distribution of WPM signals should also be chi-squared. This fact is corroborated in Figure 5.3, where the curves for the power distribution of WPM signal are plotted along with Gaussian, Rayleigh and chi-square distributions. In Figure 5.4 the power distributions for WPM signals applying different wavelet families are shown. Almost all the wavelet families have a power distribution that is chi-squared. The specifications of

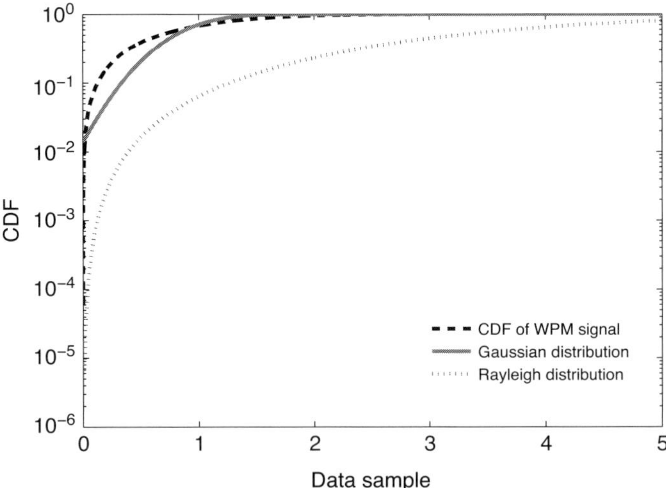

Figure 5.2 CDF of WPM signals. The wavelet considered is Daubechies of length 15; the WPM system has 128 carriers. Gaussian and Rayleigh signals are also plotted for reference.

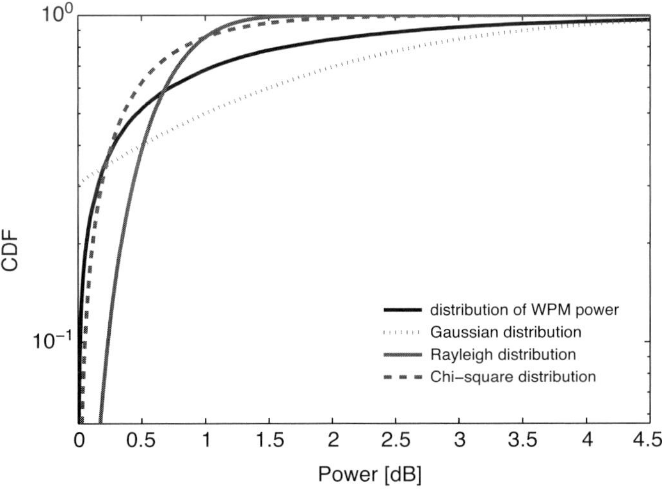

Figure 5.3 CDF of power of Gaussian and wavelet packet modulation signals. The wavelet considered is Daubechies of length 15; the WPM system has 128 carriers.

the wavelets (Daubechies 15, Coiflet 5, Symlet 15, discrete Meyer (of length 102) and Biorthogonal 2.2) that are considered are given in Table 5.1.

5.3.2.4 PAPR distribution

Figure 5.5 and Figure 5.6 show the PAPR performance curves for various wavelet families and various filter lengths, respectively. From Figure 5.5 we can deduce that apart from the biorthogonal wavelet, all the other wavelets follow a similar CDF pattern

Table 5.1 Wavelet specifications

Name of wavelet	Orthogonal	Length
Daubechies	Yes	30
Coiflet	Yes	10
Symlet	Yes	30
Discrete Meyer	Yes	102
Biorthogonal	No	(2,2)

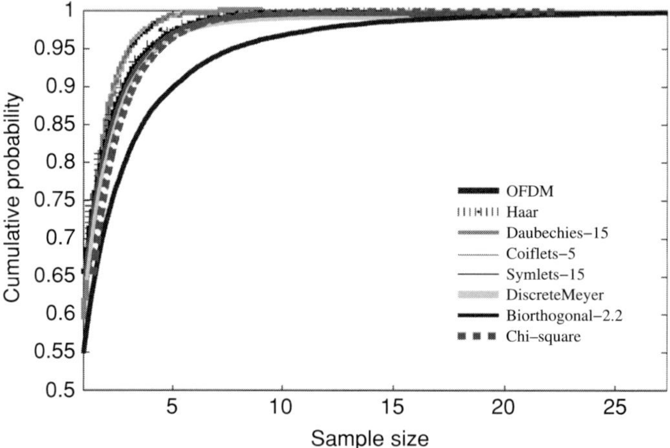

Figure 5.4 CDF of power of wavelet packet modulated signals for various families (for 128 carriers).

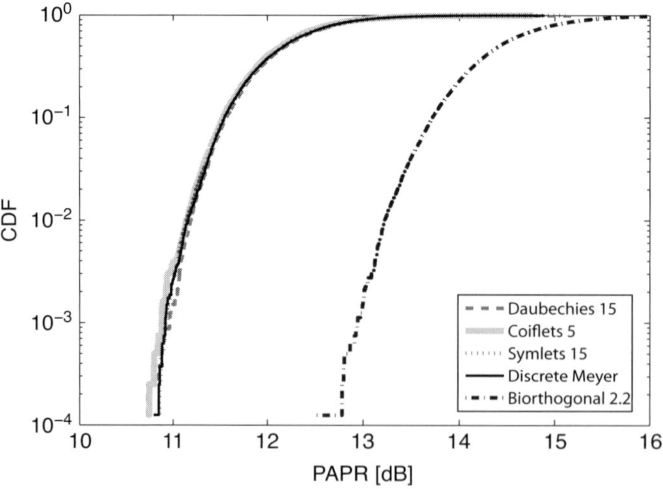

Figure 5.5 CDF of PAPR for the WPM system applying several wavelet families. All the configurations are taken to have 128 carriers.

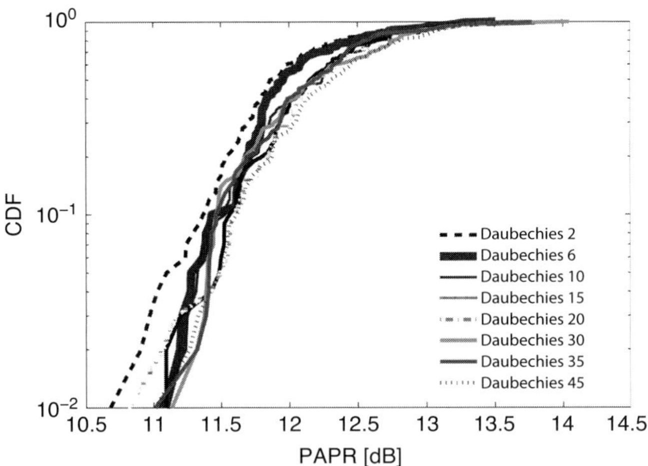

Figure 5.6 CDF of PAPR for the WPM system applying different filter lengths of the Daubechies wavelet family. The number of sub-carriers considered is 128.

for the PAPR. And from Figure 5.6 it is clear that even with increasing lengths of the wavelet, from Daubechies 2 to Daubechies 45, the PAPR distribution does not vary much.

5.4 PAPR reduction techniques

In the literature, several techniques have been proposed to reduce the PAPR of multi-carrier modulation signal. These methods can be broadly categorized into signal-scrambling techniques and signal-distortion techniques. Signal-scrambling techniques are all variations on how to scramble the codes or on how to modify phases to decrease the PAPR. Coding techniques can be used for signal scrambling. Golay complementary sequences, Shapiro–Rudin sequences and Barker codes can be used efficiently to reduce the PAPR. However, with the increase in the number of carriers the overhead associated with an exhaustive search of the best code would increase exponentially. More practical solutions of the signal-scrambling techniques are block coding, selected mapping (SLM) and partial transmit sequences (PTS). In the signal-distortion techniques the high peaks are directly reduced by distorting the signal prior to amplification. Clipping the OFDM signal before amplification is a simple method to limit PAPR. However, clipping may cause large out-of-band (OOB) and in-band interference, which results in the system performance degradation. More practical solutions are peak windowing, peak cancellation, peak power suppression, weighted multi-carrier transmission, companding, etc.

Figure 5.7 shows the classification of various PAPR reduction techniques.

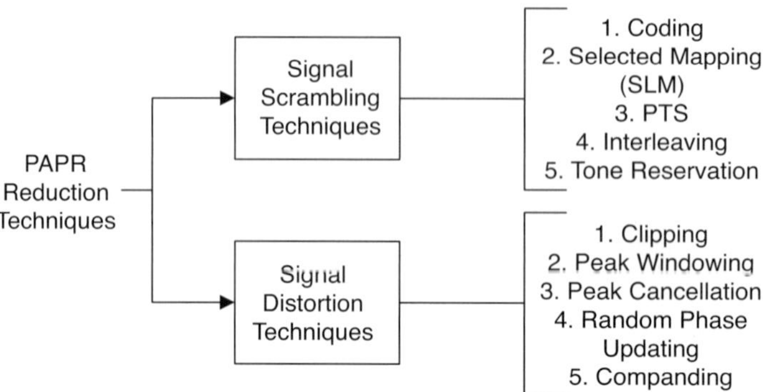

Figure 5.7 Classification of commonly known PAPR reduction techniques.

5.4.1 Signal-scrambling techniques

5.4.1.1 Block coding

Jones et al. [10] proposed a block coding scheme for the reduction of the peak to mean envelope power ratio of multi-carrier transmission systems. The main idea behind this approach is that PAPR can be reduced by block coding the data such that a set of permissible code words does not contain those that result in excessive peak envelope powers (PEPs). There are three stages in the development of the block-coding technique. The first stage is the selection of suitable sets of code words for any number of carriers, any M-ary phase-modulation scheme and any coding rate. The second stage is the selection of the sets of code words that enable efficient implementation of the encoding/decoding. The third stage is the selection of sets of code words that also offer error deduction and correction potential.

5.4.1.2 Selected mapping (SLM)

Bauml et al. [11] proposed a method for the reduction of peak to average transmit power of multi-carrier modulation systems with selected mapping. In the selected mapping (SLM) method a whole set of candidate signals is generated representing the same information, and then the most favourable signal as regards to minimum PAPR is chosen and transmitted. Information about this choice needs to be explicitly transmitted along with the chosen candidate signal. The SLM scheme is one of the initial probabilistic approaches for reducing the PAPR problem, with a goal of making the occurrence of the peaks less frequent, not to eliminate the peaks. The scheme can handle any number of sub-carriers and the drawback associated with the scheme is the overhead of side information that needs to be transmitted to the receiver.

5.4.1.3 Partial transmit sequences (PTS)

Muller and Huber [12] proposed an effective and flexible peak power reduction scheme for OFDM system by combining partial transmit sequences (PTS). The main idea behind the scheme is that the data block is partitioned into non-overlapping sub-blocks and each sub-block is rotated with a statistically independent rotation factor. The rotation factor, which generates the time-domain data with the lowest peak amplitude, is also transmitted to the receiver as side information. The PTS scheme can be interpreted as a structurally modified case of the SLM scheme, and is found that the PTS schemes performs better than SLM schemes but is much more complex. When differential modulation is used in each sub-block, no side information needs to be transmitted to the receiver.

5.4.1.4 Interleaving

Jayalath and Tellambura [13] present an interleave-based technique for improving the peak-to-average power ratio of an OFDM signal. Highly correlated data frames have large PAPR, which could thus be reduced if long correlation patterns were broken down. The paper proposes a data randomization technique for the reduction of the PAPR of the OFDM system. The most important aspect of this method is that it is less complex than the PTS method but achieves comparable results. This method is most effective for data frames with moderate PAPR values (very high PAPR values that are nearly M can not be reduced by this method). Therefore, a higher-order error-correction method should be used in addition to this scheme.

5.4.1.5 Tone reservation

The tone reservation method in [14] is suggested to reduce the PAPR. In this method a fraction of the bandwidth is used to synthesize signals that are of opposite polarity and shape a peak in the OFDM signal. Subtraction of peaks reduces the PAPR without great effect on the transmission capability of OFDM. The basic idea is to reserve a small set of tones for PAPR reduction. The problem of computing the values for these reserved tones that minimize the PAPR can be formulated as a convex problem and can be solved exactly. The amount of PAPR reduction depends on the numbers of reserved tones, their location within the frequency vector and the amount of complexity.

5.4.2 Signal-distortion techniques

5.4.2.1 Clipping

The simplest way to reduce the PAPR is the envelope clipping [15] such that the peak amplitude becomes limited to some threshold. Nevertheless, by distorting the OFDM signal amplitude, a kind of self-interference is introduced that degrades the BER. Furthermore, nonlinear distortion of the OFDM signal significantly increases the level of both in-band distortion and out-of-band radiation.

5.4.2.2 Peak windowing

Van Nee and de Wild [16] infer that since large PAPR ratios occur only infrequently, it is possible to remove these peaks at the cost of a small amount of self-interference. Clipping is one example of a PAPR reduction technique creating self-interference. The peak windowing technique provides better PAPR reduction with better spectral properties than clipping. In the windowing technique a large signal peak is multiplied with a certain window, such as a Gaussian-shaped window, cosine, or Kaiser and Hamming window. Since the OFDM signal is multiplied with several of these windows, the resulting spectrum is a convolution of the original OFDM spectrum with the spectrum of the applied window.

5.4.2.3 Peak cancellation

The peak-cancellation method introduced in [17] suggests subtracting a time-shifted and a scaled reference function from the signal, to reduce the peak power of the signal. By selecting an appropriate reference function with approximately the same bandwidth as the transmitted signal, it can be assured that the peak power reduction does not cause any out-of-band interference. BER performance results of the method as a function of the input back-off of the amplifier over AWGN and frequency-selective fading channels are reported in [17].

5.4.2.4 Random phase updating

Nikookar and Lidsheim [18] proposed a novel random phase updating algorithm for the peak to average power ratio (PAPR) reduction of the OFDM signal. In the random phase updating algorithm, a random phase is generated and assigned to each carrier. The random phase update is continued till the peak value of the OFDM signal is below the threshold. The threshold can be dynamic and the number of iterations for the random phase update is limited. After each phase update, the PAPR is calculated and the iteration is continued till the minimum threshold level is achieved or the maximum number of iterations has been reached.

5.4.2.5 Companding

Wang et al. [19] proposed a simple and effective companding technique to reduce the PAPR of OFDM signal. The OFDM signal can be assumed to be Gaussian distributed, and the large OFDM signal occurs infrequently. So, the companding technique can be used to improve OFDM transmission performance. This technique is used to compand the OFDM signal before it is converted into an analogue waveform. The OFDM signal after taking IFFT is companded and quantized. The companding technique improves the quantization resolution of small signals at the price of the reduction of the resolution of large signals, since small signals occur more frequently than large ones.

5.4.3 Criteria for selection of PAPR reduction technique

Many factors have to be considered before the right PAPR reduction technique can be chosen. Some of the factors include PAPR reduction capability, power increase in the transmit signal, BER increase at the receiver, loss in data rate, computation complexity increase, etc. Some of these requirements are contradictory and cannot be met at the same time. For example, the amplitude clipping technique clearly removes the time-domain signal peaks but results in in-band distortion and out-of-band radiation. Some techniques require a power increase in the transmitted signal after using PAPR reduction techniques. For example, tone reservation requires more signal power because some of its power must be used for the peak-reduction carriers. Some techniques may have an increase in BER at the receiver if the transmit signal power is fixed or equivalently may require large transmit power to maintain the BER after applying the PAPR reduction techniques. In some techniques such as SLM, PTS and interleaving, the entire data block may be lost if the side information is received in error. This also may increase BER at the receiver. Some techniques require the data rate to be reduced. In the block-coding technique one out of four information symbols is to be dedicated to controlling PAPR. In SLM, PTS and interleaving, the data rate is reduced due to the side information used to inform the receiver of what has been done in the transmitter. In these techniques the side information may be received in error, unless some form of protection such as channel coding is employed. Computational complexity is yet another important consideration in choosing a PAPR reduction technique. Techniques such as PTS find a solution for the PAPR reduced signal by using many iterations. The PAPR reduction capability of the interleaving technique is better for large number of interleavers, which on the other hand slows the computation capacity. Based on the above discussion, in this work we chose methods based on selected mapping for mitigation of PAPR in WPM transmission. The SLM technique is a simple and reliable technique that can be readily applied to WPM to achieve better PAPR performances.

5.5 Selected mapping with phase modification

In this section we present a method to reduce the peak-to-average power ratio (PAPR) in the wavelet packet modulator. The method works on the principle that the PAPR of a multi-carrier system can be adjusted by varying the phase shifts of the sub-carriers. Different PAPR values for the same information are obtained by randomly altering the phases of the sub-carriers used to modulate the data. The WPM carriers are randomly rotated with a value chosen from an alphabet of a finite number of identically spaced phase shifts. The WPM frame with the least PAPR is then identified and transmitted. The attraction of the methods is in its simplicity of implementation and the notable gains it yields with minimal increase in complexity.

5.5.1 Description of algorithm

Figure 5.8 shows the blocks of the proposed WPM system with the PAPR reduction structure. The bit stream from the information source is first converted to a constellation QPSK/BPSK stream and then replicated to obtain a finite number of copies, say L. Each of the replicated set is then serial-to-parallel (S/P) converted and then phase shifted by a random phase sequence.

The phase sequences are generated by a phase generator that chooses between different phase alphabets ϕ and stochastic distributions and creates a phase vector $\Phi_p^{(n)}$. Here, $n = 1, 2, 3, \ldots, L$ stands for the index of the frame and $p = 1, 2, 3, \ldots, M$ connotes the sub-carrier index. The phase vector thus contains L rows each with M columns. Denoting

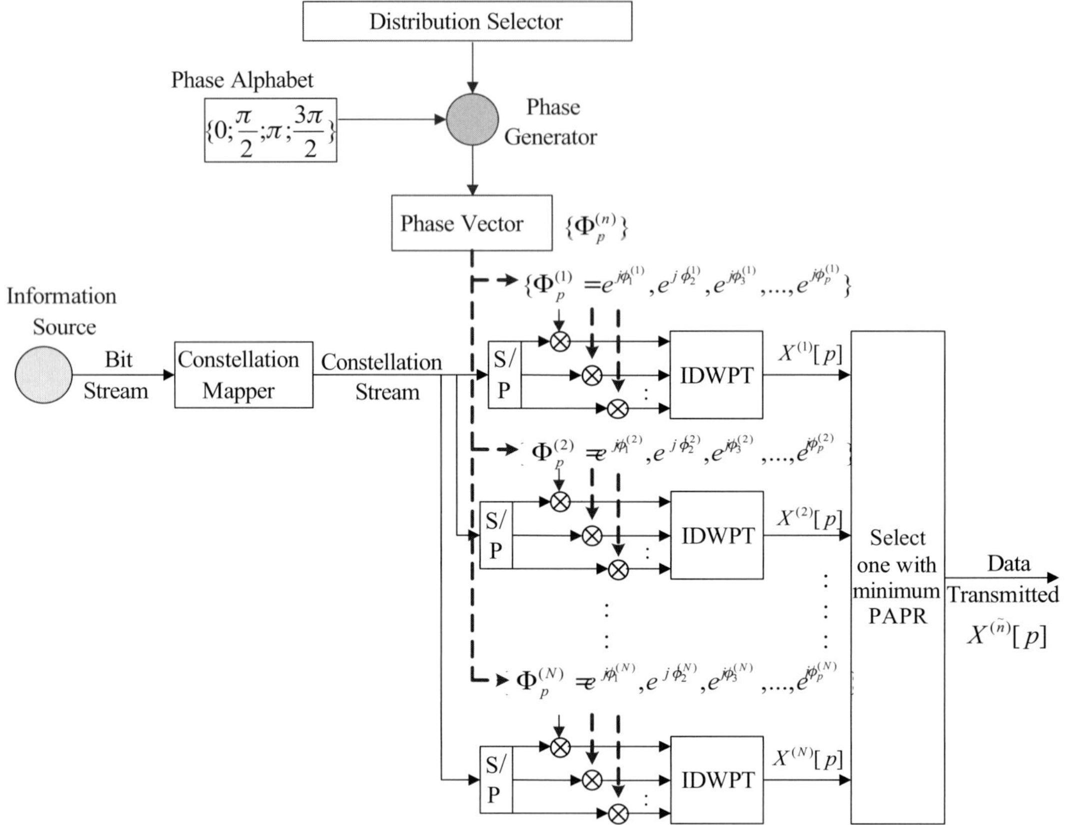

Figure 5.8 WPM transmitter block diagram with the SLM-based PAPR reduction technique.

the information-bearing WPM frame by the notation $\mathbf{X}[p]$, the L different WPM frames $\mathbf{X}^{(n)}[p]$ obtained by sub-carrier-wise multiplication with the phase vector $\Phi_p^{(n)}$

$$\mathbf{X}^{(n)}[p] = \mathbf{X}[p] \times \Phi_p^{(n)} = \mathbf{X}[p] \times e^{j\phi_p^{(n)}} \tag{5.6}$$

The phase-shifted information bearing streams are then transformed by an IDWPT operation and the PAPR of the transformed composite signal is calculated. Amongst the set of L PAPR values, the one with the lowest value is selected and transmitted. Note that all of the frames carry identical information. In order to achieve PAPR reduction, the WPM frame with the lowest PAPR is transmitted. Defining the candidate time-domain WPM frame as $x = \text{IDWPT}(\mathbf{X}^{(n)}[p])$, the index of this frame can be given as:

$$\hat{n} = \underset{1 \leq n \leq L}{\arg\min}(\text{PAPR}(x^{(n)})) \tag{5.7}$$

In order to ensure that the transmitter and receiver operate harmoniously, the chosen index of the frame $\hat{l}$ is sent to the receiver as a side information. Typically for a size L vector, the number of bits required to send $\hat{l}$ will be $\log_2(L)$. However, to prevent corruption of this precious message, more bits may be used to encapsulate this message by channel coding.

The algorithm to calculate and select the minimum PAPR for WPM is summarized in Algorithm 1.

Algorithm 1: Pseudocode for calculating and selecting the minimum PAPR for WPM

1: Obtain the source message.
2: Replicate it a finite number of times, say L.
3: Generate statistically independent phase sequences from the chosen phase alphabet (e.g., $\phi \in (0, \pi/2, \pi, 3\pi/2)$).
4: Multiply frame sequences element/carrier-wise by M-length phase sequences. Here, M is also the number of WPM carriers.
5: Do the IDWPT transform for each resulted frame sequence for each replicated copy of the data.
6: Calculate the PAPR per frame of the signal for each replicated copy of the data and find the PAPR.
7: List all the PAPR values; select the minimum PAPR and transmit.
8: Send as side information the index of the frame with minimum PAPR, $\hat{l}$ to recover the data in the receiver.

5.5.2 Numerical results

In this section we present results of the studies and evaluate the performance of WPM system with the PAPR reduction technique. The investigations are carried out using computer simulations and the performance metric of choice is the Complementary CDF (i.e., CCDF). The WPM system is realized using a filter bank structure with 7 levels of decomposition. The modulation scheme used is QPSK. The phase alphabet is taken to be $\phi \in (0, \pi/2, \pi, 3\pi/2)$, which is randomly chosen while generating the phase vector. The wavelet of choice is Daubechies 5 (denoted db5), which is of length 10. These simulation parameters will be used throughout the experiments unless stated otherwise. To properly evaluate the improvements due to the PAPR reduction technique, a reference PAPR-CCDF curve obtained for a Daubechies 5 (db5) wavelet for the case without PAPR reduction (i.e., no phase modification) will also be provided.

5.5.2.1 Performance of the PAPR mitigation technique

In the results we verify the impact of the PAPR reduction technique. Figure 5.9 shows the CCDF curves for the variation of PAPR under the PAPR reduction technique for different number of replications, L. A reference curve with no PAPR reduction is also plotted. It is evident from the plots that the improvements are significant and bring in up to 3 dB reduction in PAPR in comparison to the case when no PAPR reduction technique is used.

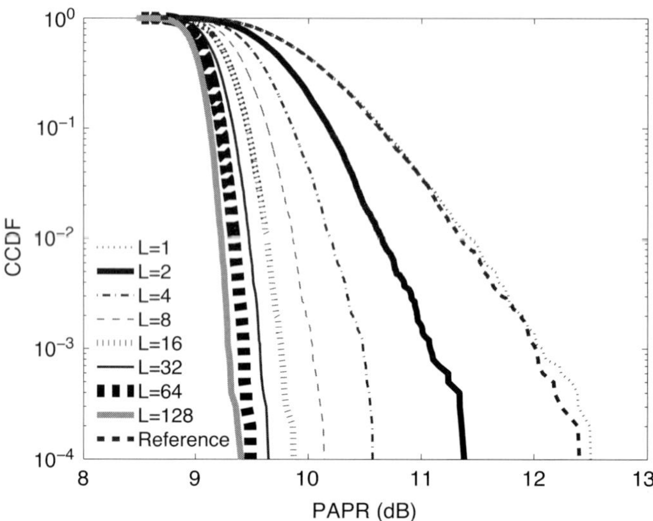

Figure 5.9 CCDF of the PAPR of the WPM signal for different values of L. The wavelet considered is Daubechies 5 (length 10). A reference curve with no PAPR reduction is also plotted.

5.5.2.2 Influence of phase sequence distribution

To gauge the impact of the distribution of the phase sequences, we now consider different stochastic distributions, namely, the random sequences, the Golay sequences [20], [21] and the Hadamard sequences where the number and length of all the sequences are the same. The phase alphabet is taken to be $\phi \in (0, \pi/2, \pi, 3\pi/2)$ and the value of L is fixed at 8. Figure 5.10 shows the respective plots and it can be deduced from the figures that though all the distributions yield notable improvements, there is no perceivable difference in their performances.

These results are important because the similarity in the performances when using pseudorandom and random sequences means that the receiver only has to know the key used to generate the pseudorandom phase sequences used at the transmitter (instead of the entire phase sequence). This aids in producing a significant reduction of the side information.

5.5.2.3 Impact of phase alphabet

We now evaluate the impact of the phase alphabet on the PAPR reduction mechanism. The results are plotted in Figure 5.11 where a range of cardinalities for the phases are considered. The results show that the choice of the phase alphabet does not affect the performance of the PAPR reduction technique.

5.5.2.4 Impact of wavelet families

We now analyze the conduct of the PAPR reduction technique for different wavelet families and for different filter lengths. In this set of experiments the value of L is taken to be 8. The various wavelet families considered are Daubechies 15, Coiflet 5, Symlet

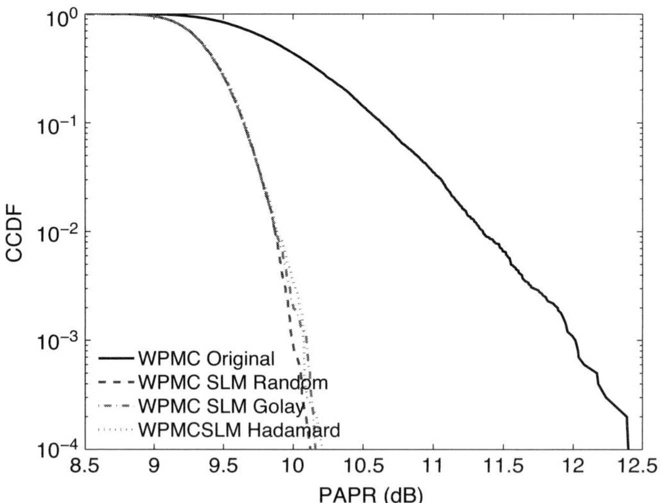

Figure 5.10 Complementary cumulative distribution function CCDF of the PAPR of WPM for different distributions of the phase sequences. The wavelet considered is Daubechies 5 (length 10).

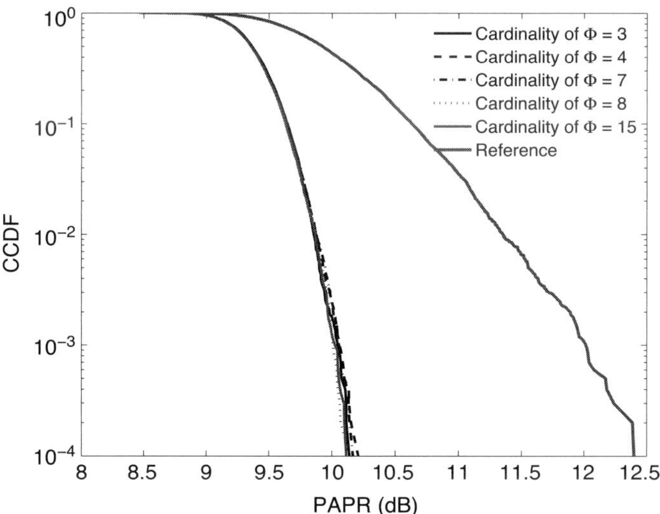

Figure 5.11 CCDF of the PAPR of WPM using the PAPR reduction technique for different phase sequences. The wavelet considered is Daubechies 5 (length 10).

15 (all of length 30), Meyer (of length 102) and Haar. Figure 5.12 and Figure 5.13 show the PAPR performance curves for various wavelet families and various filter lengths, respectively. From Figure 5.12 we can deduce that all the wavelets follow a similar CCDF pattern for their PAPR performances. And from Figure 5.13 it is clear that even with increasing lengths of the wavelet filter, from Daubechies 2 to Daubechies 35, the

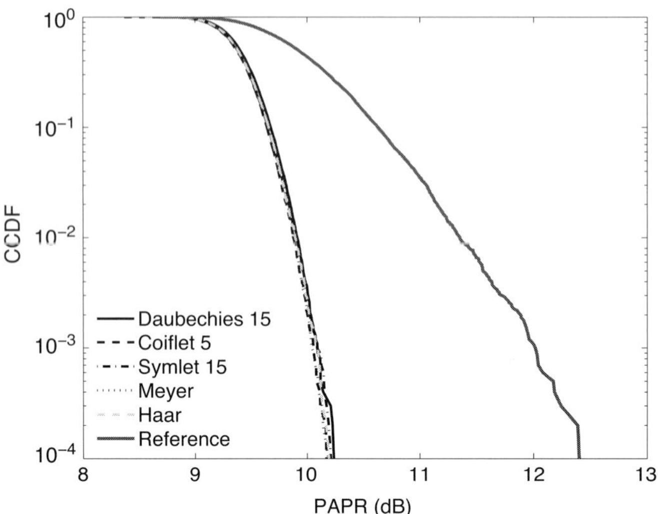

Figure 5.12 CCDF of the PAPR for several wavelets.

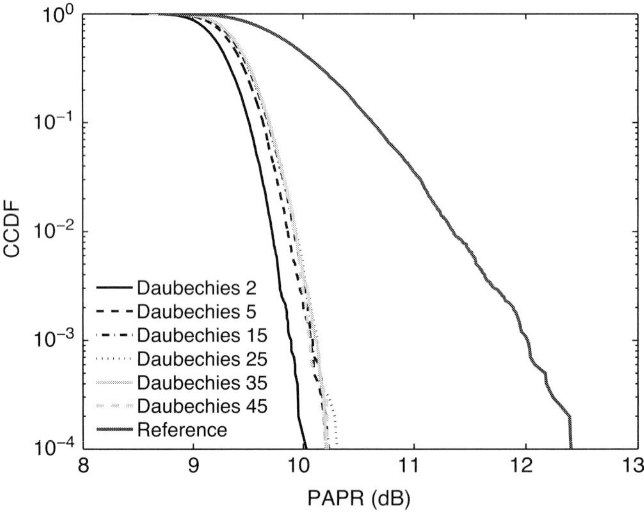

Figure 5.13 CCDF of the PAPR for the WPM system with different filter lengths of the Daubechies wavelet family.

PAPR distribution is limited to a variation of about 0.8 dB. In all instances the proposed technique reduces the PAPR between 1.5 and 2.5 dB.

5.5.2.5 Influence of the PAPR reduction technique on the BER performance

We finally plot the BER performances of the WPM system (Figure 5.14). The curves plotted are for the cases when the phase sequences are generated randomly (two cases are considered where in the first case the receiver has complete and perfect knowledge of the random phases used at the transmitter and in the second case where the receiver

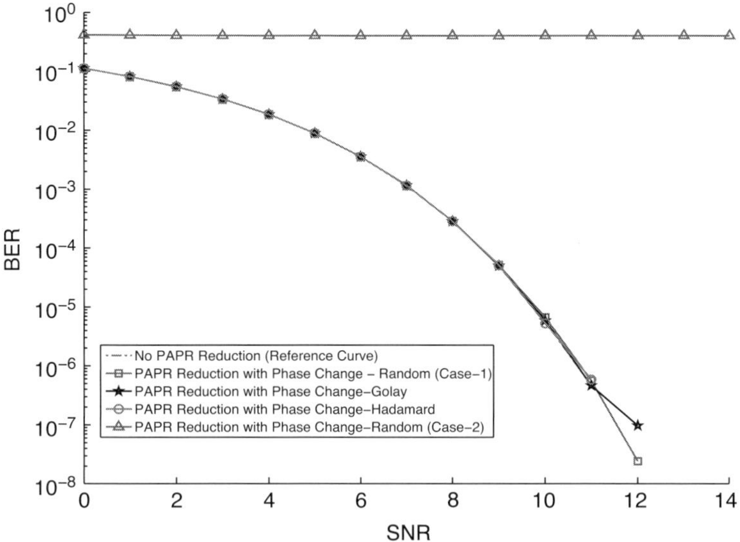

Figure 5.14 BER vs. SNR (dB) for the cases with and without PAPR reduction (phase modification) and for various distributions of the phase. For the case with random phase change, two figures are plotted. In the first case (denoted Case-1) the receiver has complete and perfect knowledge of the random phases used at the transmitter. In the second scenario (marked Case-2) the receiver operates with no knowledge of the phases used at the transmitter. For the scenarios when phases with Golay and Hadamard distributions are used, the transmitter and receiver only share the keys of the pseudorandom polynomial.

operates with no knowledge of the phases used at the transmitter) and pseudorandomly (when the receiver only knows the key used by the transmitter to generate the pseudorandom sequence). As a reference, the BER plot for the case with no PAPR reduction technique is also plotted. The results show the importance of having complete knowledge on the phase sequences. Even a slight mismatch in the phase information at the receiver degrades the system performance significantly[1]. Since a perfect replication of randomly generated phases is not possible at the receiver, the application of pseudorandom generators can be considered. This is supported by the results plotted in Figure 5.10 where the PAPR reduction due to pseudorandom codes is shown to be as good as that of random phase generators and in Figure 5.14 where the BER curves show that using a PAPR reduction mechanism with pseudorandom phase generators does not result in any loss in performance.

5.6 Summary

In this chapter, a study on the effect of PAPR on the developmental wavelet packet modulator was presented. The stochastic distribution of the WPM signal and its power variations were analyzed. The studies showed that the envelope of the WPM signal is

[1] As we saw in Chapter 4, multi-carrier systems like OFDM and WPM are sensitive to phase errors.

Gaussian and its power has a chi-squared distribution. The PAPR for WPM systems with different wavelets, pulse shapes and lengths were studied. Furthermore, strategies to mitigate the PAPR in the WPM systems were proposed.

References

[1] B. Torun, M. Lakshmanan, and H. Nikookar, "On the Analysis of Peak-to-average Power Ratio of Wavelet Packet Modulation," in *Proc. of European Wireless Technology Conference*, Rome, Italy, 29 Sept.–1 Oct. 2009, pp. 1–4.

[2] M. Baro and J. Ilow, "Improved PAPR Reduction for Wavelet Packet Modulation using Multi-pass Tree Pruning," in *Proc. of the IEEE 18th International Symposium on Personal, Indoor and Mobile Radio Communications, PIMRC*, Athens, Greece, 3–7 Sept. 2007, pp. 1–5.

[3] H. Zhang, D. Yudo, and F. Zhao, "Research of PAPR Reduction Method in Multicarrier Modulation System," in *Proc. of the International Conference on Communications, Circuits and Systems*, vol. 1, Hong Kong, China, 27–30 May 2005, pp. 91–4.

[4] N. Le, S. Muruganathan, and A. Sesay, "Peak-to-average Power Ratio Reduction for Wavelet Packet Modulation Schemes via Basis Function Design," in *Proc. of the IEEE 68th Vehicular Technology Conference*, VTC 2008-Fall, Calgary, BC, Canada, 21–24 Sept. 2008, pp. 1–5.

[5] V. Kumbasar and O. Kucur, *Digital Signal Processing*, Elsevier, Nov. 2008, vol. 18, no. 6, ch. Better wavelet packet tree structures for PAPR reduction in WOFDM systems, pp. 885–91.

[6] R. Van Nee and R. Prasad, *OFDM for Wireless Multimedia Communications*, Artech House, Inc. Norwood, MA, 2000.

[7] S. Litsyn, *Peak Power Control in Multicarrier Communications*, Cambridge University Press, 2007.

[8] M. Gautier, C. Lereau, M. Arndt, and J. Lienard, "PAPR Analysis in Wavelet Packet Modulation," in ISCCSP 2008, *3rd International Symposium on Communications, Control and Signal Processing*, 2008, St Julians, Portugal, 12–14 Mar. 2008, pp. 799–803.

[9] A. Jamin and P. Mahonen, "Wavelet Packet Modulation for Wireless Communications," *Wireless Communications and Mobile Computing Journal*, vol. 5, no. 2, pp. 123–37, Mar. 2005.

[10] A. Jones, T. Wilkinson, and S. Barton, "Block Coding Scheme for Reduction of Peak to Mean Envelope Power Ratio of Multicarrier Transmission Schemes," *Electronics Letters*, vol. 30, p. 2098, 1994.

[11] R. Bauml, R. Fischer, and J. Huber, "Reducing the Peak-to-average Power Ratio of Multicarrier Modulation by Selected Mapping," *Electronics Letters*, vol. 32, pp. 2056–7, 1996.

[12] S. Muller and J. Huber, "OFDM with Reduced Peak-to-average Power Ratio by Optimum Combination of Partial Transmit Sequences," *Electronics Letters*, vol. 33, no. 5, 1997.

[13] A. Jayalath and C. Tellambura, "The Use of Interleaving to Reduce the Peak-to-average Power Ratio of an OFDM Signal," in *IEEE Global Telecommunications Conference, GLOBECOM* 2000, vol. 1, 2000, pp. 82–6.

[14] J. Tellado, *Multicarrier Modulation with low PAPR: applications to DSL and Wireless*, Springer Netherlands, 2000.

[15] D. Kim and G. Stuber, "Clipping Noise Mitigation for OFDM by Decision-aided Reconstruction," *IEEE Communications Letters*, vol. 3, no. 1, pp. 4–6, 1999.
[16] R. van Nee and A. de Wild, "Reducing the Peak-to-average Power Ratio of OFDM," in *Proc. of the 48th IEEE Vehicular Technology Conference, VTC 98*, vol. 3, 1998, pp. 2072–6.
[17] T. May and H. Rohling, "Reducing the Peak-to-average Power Ratio in OFDM Radio Transmission Systems," *Proc. IEEE Vehicular Technology Conference, VTC 98*, pp. 2774–8, 1998.
[18] H. Nikookar and K. Lidsheim, "Random Phase Updating Algorithm for OFDM Transmission with Low PAPR," *IEEE Transactions on Broadcasting*, vol. 48, no. 2, pp. 123–8, 2002.
[19] X. Wang, T. Tjhung, and C. Ng, "Reduction of Peak-to-average Power Ratio of OFDM Systems using a Companding Technique," *IEEE Transactions on Broadcasting*, vol. 45, no. 3, 1999.
[20] J. Davis and J. Jedwab, "Peak-to-mean Power Control and Error Correction for OFDM Transmission using Golay Sequences and Reed–Muller Codes," *Electronics Letters*, vol. 33, no. 4, pp. 267–8, Feb. 1997.
[21] J. Davis and J. Jedwab, "Peak-to-mean Power Control in OFDM, Golay Complementary Sequences, and Reed–Muller Codes," *IEEE Transactions on Information Theory*, vol. 45, no. 7, p. 2397–2417, 1999.

6 Wavelets for spectrum sensing in cognitive radio applications

6.1 Background

Cognitive radio (CR) is an emerging technology for the strategic utilization of the radio spectrum, one of the most valuable resources for any wireless application. Fast and accurate sensing of the radio spectrum used for wireless applications is one of the key requirements of any cognitive radio architecture. From the survey of all traditional and recently developed spectrum sensing methodologies, it is observed that a low complexity reconfigurable spectrum estimator with the property of a flexible time–frequency resolution is desirable. A recently developed technique for the detection of active users or spectrum holes in the spectral domain is the wavelet-packet-based spectrum estimator (WPSE). In this chapter, first in Section 6.2, the general idea of spectrum sensing in cognitive radio is briefly explained followed by the description of methods of spectrum sensing in Section 6.2. The advantages and disadvantages of conventional spectrum sensing techniques are studied in Section 6.4 and for wavelets in Section 6.5. The receiver operating characteristics (ROC) criteria are selected for the evaluation of the performance of wavelet-packet based spectrum estimator in Section 6.6, and accordingly the variation of the ROC performance is studied in Section 6.7 for different wavelet families (both orthogonal and non-orthogonal), wavelet filters, number of primary users and different signal-to-noise ratios of the input signal. Exploiting the prominent underutilization of the radio spectrum, an efficient wavelet-packet based spectrum estimation model with reduced number of sensing measurements is proposed in Section 6.8 where a significant reduction of number of sensing measurements with a promising detection performance of the developed model is evaluated by extensive simulations. Furthermore, in Section 6.9 the establishment of WPSE in the light of theory of compressive sensing is highlighted. And finally, a summary of the chapter appears in Section 6.10.

6.2 Spectrum sensing in cognitive radio

Cognitive radio (CR) [1], [2] is a revolutionary technology that enables the efficient spectrum utilization by allowing the unlicensed users to transmit in licensed user bands when they are vacant. Primary users (PU)/licensed users are the actual owners of a fixed spectrum and secondary/cognitive users opportunistically use the licensed user spectrum when the primary users are inactive.

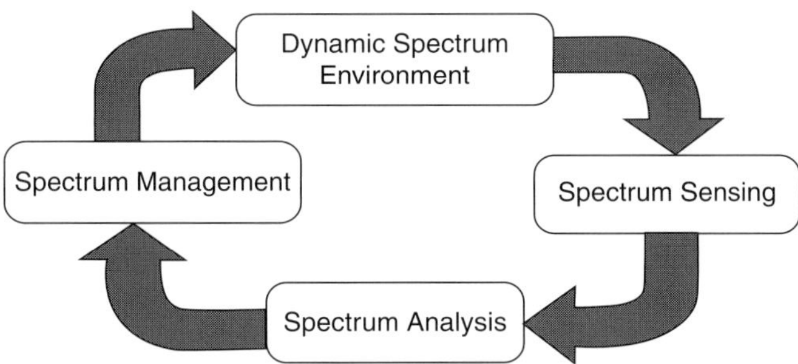

Figure 6.1 Simplified block diagram of cognitive radio (CR) cycle.

Rapid growth of wireless communications and its applications creates scarcity of the radio spectrum, which is the most valuable resource in wireless technology [3]. But at the same time significant underutilization of the radio spectrum by licensed wireless users at any particular time and geographic locations is also recognized [4], [5]. Cognitive radio is an intelligent wireless technology that is aware of the variation of the dynamic spectrum environment and adapts its transmission parameters in order to efficiently utilize the radio spectrum used for several wireless applications. The main architecture of any CR-based system is inspired by the basic human cognition cycle. In the cognitive-radio paradigm there are broadly three main steps that are associated with the opportunistic spectrum access. These steps are analogous to learning from the environment, taking a decision and acting according to that. The environment is the dynamic spectrum environment in a CR-based system. In Figure 6.1 a simplified block diagram of a basic cognitive radio cycle is demonstrated.

One of the main functionalities of any CR-based system is the fast and accurate sensing of the radio spectrum to estimate the locations of active/primary users or the vacant/unused portions of the spectrum known as spectrum holes. Any sensing technique of the radio spectrum must be robust against the statistical variations of the highly varying dynamic spectrum. Some fundamental issues related to any spectrum sensing methodology for CR applications are summarized below:

1. *Speed of the estimation:* The process of estimation should be fast. In other words, the time resolution of the estimate should be improved. This criterion is very important for dynamic spectrum access (DSA).
2. *Accuracy of the estimation:* The frequency resolution of the estimate defines the accuracy of the estimated spectrum. It signifies that different frequency regions and their characteristics should be estimated correctly. In this case, a flexible and reconfigurable method that can dynamically tune its frequency resolution is also desirable. Side-lobes and leakage of the estimate should also be minimized to achieve a good level of accuracy.
3. *Complexity of the method:* In any wireless application the complexity of the technology is a very important issue. The power consumption for spectrum sensing can be optimized by reducing the number of sensing measurements. Optimization of the

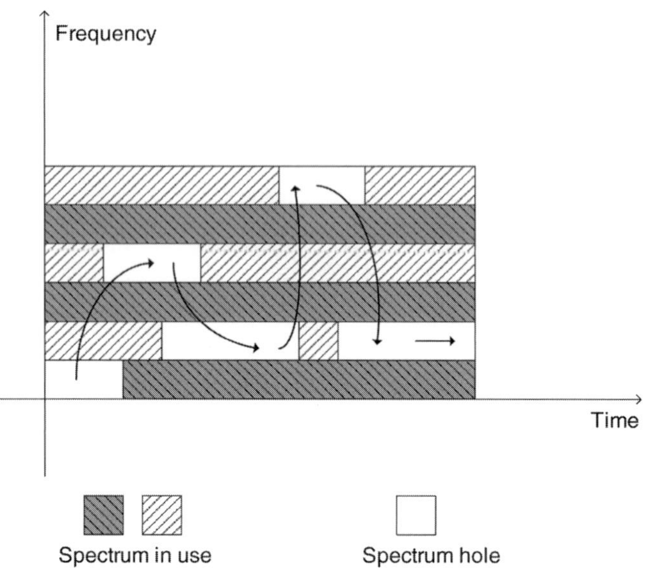

Figure 6.2 Spectrum sensing and dynamic spectrum access.

number of secondary/cognitive user nodes can also reduce the complexity and power consumption of the sensing scheme.

Figure 6.2 illustrates an example of dynamic sensing and access of any spectrum [25].

6.3 Spectrum sensing methods

6.3.1 Periodogram

One of the most important techniques of spectrum analysis is the periodogram. This is a non-parametric approach to estimate the power spectral density (PSD) of any spectrum of random signals (which does not rely on any model-based approximations like moving average(MA), auto-regressive (AR) or auto-regressive moving average (ARMA)). The disadvantage of the parametric method is that if the signal is not sufficiently and accurately described by the model, the result is not perfect. Non-parametric methods, on the other hand, do not have any assumption about the shape and structure of the spectrum and try to find an estimate of the power spectrum without any previous knowledge about the hidden stochastic approach.

A periodogram is a direct approach as it calculates the PSD directly from the input signal. In practice, for a signal $x[n]$ with N samples, first the discrete Fourier transform (DFT) is calculated. Then the DFT is squared and its average is taken over N samples. Mathematically, it is written as,

$$\Gamma(e^{j\omega}) = \frac{1}{N}\left|X(e^{j\omega})\right|^2 = \frac{1}{N}\left|\sum_{n=0}^{N-1} x[n]\, e^{-j\omega n}\right|^2 \tag{6.1}$$

Here, $\Gamma(e^{j\omega})$ is the periodogram of the signal $x[n]$ and $X(e^{j\omega})$ is the transform in the frequency domain. Basically in the periodogram, the signal is windowed via a rectangular window sequence. There are some specific disadvantages of a periodogram that give rise to the modifications and alterations of the basic periodogram. The main disadvantages are:

- The estimates of the power spectral density (PSD) are coarse with low precision and large variance that does not improve with longer sequences. (The rectangular window results in a Dirichlet kernel described by the width of the main lobe and the level of side-lobe. The width of the main lobe is related to the frequency resolution of the power spectra, and the level of the side-lobe is related to the ratio between maximum and minimum power of the spectrum.)
- The use of the rectangular window compromises the frequency resolution, producing leakage, and the estimate is biased.

The modifications of the periodogram to improve the performance are:

1. Bartlett method, where the samples are divided into several segments and the periodogram of each segment is averaged.
2. Welch method, which is the modification of the Bartlett method by allowing the segments overlap and introducing different windows (such as Blackman, Hamming or Kaiser window) on signals before the calculation of the periodogram. This approach introduces the tuning of the resolution and the variance of the power spectra.

6.3.2 Correlogram

The indirect approach of PSD calculation is the correlogram. In a correlogram, the autocorrelation function of the input signal $\phi_{xx}[k]$ is computed. The power spectral density (PSD) is then obtained from the Fourier transform of $\phi_{xx}[k]$ illustrated as:

$$\Gamma(e^{j\omega}) = \sum_{k=-\infty}^{\infty} \phi_{xx}[k]e^{-j\omega k} \tag{6.2}$$

The calculation of the true autocorrelation value requires an infinite length of sequence but only an approximation is possible. In general, there are two possible ways to compute the approximation of autocorrelation value, i.e., $\hat{\phi}_{xx}[k]$, namely standard biased and standard unbiased estimates. An extension of the above-mentioned approach is application of a selected window function $w[k]$, and then taking the Fourier transform, to reduce the variance and smoothing the periodogram of the power estimate that is known as the Blackman–Tukey method. Mathematically, the PSD in this case is written as,

$$\Gamma(e^{j\omega}) = \sum_{k=-L_w}^{L_w} \hat{\phi}_{xx}[k].w[k].e^{-j\omega k} \tag{6.3}$$

where $(2L_w + 1)$ is the length of the selected window and L_w is generally less than the total number of samples of available data.

This smoothing of the signal plays an important role to reduce the bias of the estimated PSD, but this convolution process would reduce the frequency resolution.

6.4 Advantages and disadvantages of conventional spectrum sensing techniques in cognitive radio

The main action in cognitive radio is the scanning of the licensed spectrum by the secondary (cognitive) users to locate the unoccupied portion of the spectrum. The challenge is to implement the above-mentioned task in such a way that is efficient for a highly time-varying spectrum. The process must be equipped to detect the change of the radio stimuli in both time and space. Some advantages and disadvantages of conventional techniques are illustrated below.

6.4.1 Pilot detection via matched filtering

This process states that the primary user sends a pilot signal [6]. The pilot signal should be known by secondary users to allow them to perform timing and carrier synchronization. Secondary users should have full prior knowledge of modulation type, pulse shaping and packet format. This is considered as a coherent technique.

Advantage: Pilot detection requires minimum sensing time because it exploits available prior knowledge, especially the carrier frequency of the primary users.

Disadvantage: To implement this scheme, secondary users should provide separate dedicated receivers for each primary user class, which is impractical from a complexity point of view. Except for this, the method also suffers from susceptibility to frequency offsets and the resultant loss of synchronization.

6.4.2 Energy detection

It is a non-coherent detection technique, where prior knowledge of pilot data is not required. The method consists of a low pass pre-filter to remove noise and interference, an analogue to digital converter as well as square law device to compute the energy, all in a cascaded condition [7]. Later, the test statistics is calculated via averaging of the samples. Here, the periodogram scheme can also be incorporated by taking the average of the square magnitudes of FFT.

Advantage: No prior knowledge (need for pilot signal) is required here.

Disadvantage: As it is a non-coherent detection, the accuracy to the detection threshold to noise, in-band interference (intermodulation effects) and fading is not up to the mark.

6.4.3 Cyclostationary feature detection

In any communication system whenever the signal is modulated with sinusoid carriers, cyclic prefixes (OFDM) and code or hoping sequences (CDMA) of the statistics of

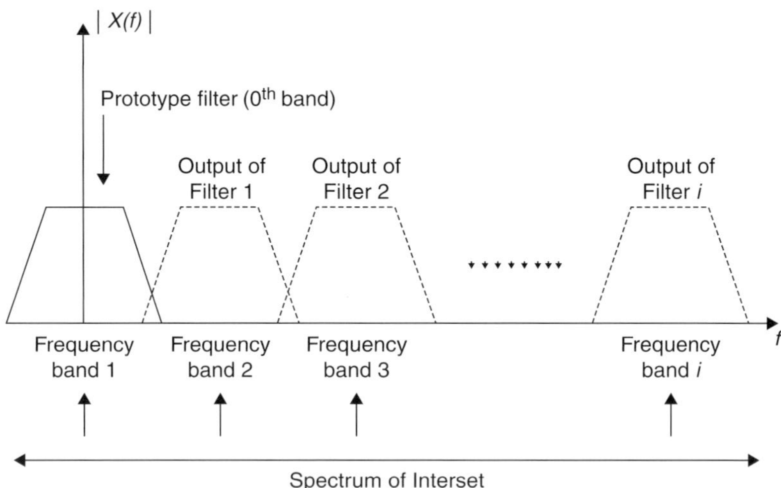

Figure 6.3 Spectrum estimation as a filter bank problem.

the signal (mean, autocorrelation, etc.) show a periodic behaviour. This property is the "cyclostationarity" of the signal and mathematically it is categorized by a function known as a spectral correlation function. The periodicity of the statistics of the signal is denoted by the cycle frequency. When the cycle frequency becomes zero, the spectral correlation function becomes the PSD of the signal.

Advantage: This method is helpful for OFDM and CDMA signals.
Disadvantage: The limitation of the method is that the transmitted data is taken to be a stationary random process.

One of the most important challenges of spectrum estimation is to reduce the side-lobe and leakages and to improve the variance. Later, it was realized that spectrum estimation can be regarded as a filter bank analysis problem. For the estimation of any spectrum, an array of bandpass filters (in parallel) can be used to separate the input signal into several frequency components, each one of a single-frequency band. The filter banks are usually implemented based on a single prototype low-pass filter. This low-pass filter is normally used to realize the zeroth band of the filter bank, while filters in the other bands are formed through modulation of the prototype filter. Figure 6.3 illustrates the idea.

The two important spectrum estimation processes, multi-taper spectrum estimation (MTSE) and filter bank spectrum estimation (FBSE), are based on the above principle.

6.4.4 Multi-taper spectrum estimation (MTSE)

The multi-taper spectrum estimator (MTSE) [8] uses multiple orthogonal prototype filters. In this process the mathematical tool of discrete prolate spheroidal sequence (also known as a Slepian sequence after the name of the inventor, David Slepian) is used. In the multi-taper method, for reducing variance the average over different tapers using the full data is implemented.

First, the last N received samples are collected in a vector given by, $x[n] = [x[n]$ $x[n-1]\ldots\ldots\ldots\ x[n-N+1]]^T$. Then, it is represented as a set of orthogonal Slepian basis vectors given by,

$$x[n] \approx \sum_{k=0}^{P-1} K_k(f_i)\bar{\mathrm{D}}\bar{\mathrm{q}}_{\mathrm{k}} \tag{6.4}$$

where $K_k(f_i)$ is the expansion coefficients and P is the number of frequency points (f_i) that have to be estimated. $\bar{\mathrm{q}}_{\mathrm{k}}$ is the set of orthogonal Slepian basis vectors and $\bar{\mathrm{D}}$ is a diagonal matrix of elements 1, $e^{j2\pi f_i}, \ldots\ldots\ldots, e^{j2\pi(N-1)f_i}$, etc. Now, the term $K_k(f_i)$ is given as

$$K_k(f_i) = (\bar{\mathrm{D}}\bar{\mathrm{q}}_{\mathrm{k}})^{\mathrm{H}} x[n] \tag{6.5}$$

Based on Eq. (6.5) the MTSE can be formulated as,

$$\hat{\Gamma}_{MTSE}(f_i) = \frac{1}{P}\sum_{k=0}^{P-1} |K_k(f_i)|^2 \tag{6.6}$$

If the vector $\bar{\mathrm{q}}_{\mathrm{k}}$ is only containing 1s, then Eq. (6.6) is the periodogram with rectangular window. If the matrix is manipulated by its elements, then windowed periodogram is achieved. In this way, $\hat{\Gamma}_{MTSE}(f_i)$ is the average of several periodograms with different types of windows.

From the viewpoint of filter bank analysis, each point f_i in the power spectrum estimate is related to the outputs of a group of bandpass filters in the same band. The pass band of the band pass filters BPF gives the frequency resolution of the system. To reduce the variance of the estimated power, the multiple prototype filters' impulse responses are derived from the $\bar{\mathrm{q}}_{\mathrm{k}}$ matrix. Here, the orthogonality of the Slepian sequences is utilized. The autocorrelation matrix for the vector $x[n]$ is calculated and then the N eigenvalues and corresponding N eigenvectors are calculated through several optimizations. Rather, P eigenvectors are chosen from N eigenvectors that give largest eigenvalues. These P prototype filters are expected to have minimum energy in the stop-band. An iterative algorithm is proposed in [8] to calculate an efficient estimation of the spectrum taking care of above constraints.

One of the main objectives of this approach is effective reduction of estimated power variance. This is precisely taken care of because of the independence of the output of each BPF.

Advantage: MTSE is an accurate method for spectrum estimation because of its tunable filter coefficients (use of multiple prototype filters) using the help of a prolate sequence.

Disadvantage: In any given frequency band of interest MTSE uses multiple prototype filters, which introduces complexity in wideband spectrum sensing in cognitive radio.

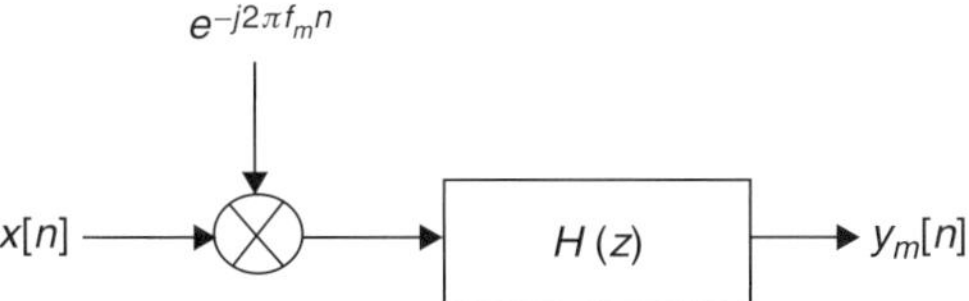

Figure 6.4 Demodulation with mth sub-carrier before processing through the root-Nyquist filter.

6.4.5 Filter bank spectrum estimation (FBSE)

FBSE is a scheme for spectrum sensing for cognitive radio incorporating multi-carrier modulation techniques. The main difference of this method from MTSE is that in FBSE, only one prototype filter is used for any particular band of interest. (In MTSE, multiple prototype filters are used for every point in the spectrum for a given frequency band.)

The filters at the receiver and the transmitter side are a pair of matched root-Nyquist filters [9]. The received signal is demodulated by the mth sub-carrier and then forwarded to the root-Nyquist filter.

Considering Figure 6.4, it can be shown that

$$S_{y_m y_m}(f) = S_{xx}(f + f_m)|H(e^{j2\pi f})|^2 \tag{6.7}$$

$S_{y_m y_m}(f)$ is the PSD of $y_m[n]$. Assuming that $H(z)$ is narrowband, $S_{xx}(f + f_m)$ can be considered as $S_{xx}(f_m)$. Now, taking the Z-transform of both sides, it can be written as

$$\Phi_{y_m y_m}(z) = S_{xx}(f_m)H(z)H(z^{-1}) \tag{6.8}$$

$S_{xx}(f_m)$ in Eq. (6.8) is a constant. $H(z)$ is designed as a root-Nyquist filter offering zero crossing in the filter in an interval of N samples. So the Nyquist filter is $G(z) = H(z)H(z^{-1})$ of order N. So, it is clear that the autocorrelation sequence $\phi_{y_m y_m}(k)$, can be derived from the inverse Z-transform of Eq. (6.8). The transfer function of the Nyquist filter should be adjusted in a way that in the time domain,

$$g[n] = \begin{cases} 1, & \text{if } n = 0 \\ 0, & \text{if } n = kN \end{cases} \tag{6.9}$$

where k is any non-zero integer. The output estimate is constructed as an observation vector of length K and having sample spacing of L_s samples.

$$y_m[n] = [y_m[n], y_m[n - L_s], \ldots\ldots\ldots\ldots, y_m[n - (K - 1)L_s] \tag{6.10}$$

The autocorrelation matrix is given by $\overline{\mathrm{R}_{y_m y_m}} = S_{xx}(f_m)\overline{\mathrm{A}}$ with

$$\overline{\mathrm{A}} = \begin{bmatrix} g_N(0)\ldots\ldots\ldots\ldots & g_N(L_s)\ldots\ldots\ldots\ldots & g_N((K-1)L_s) \\ g_N(-L_s)\ldots\ldots\ldots & g_N(0)\ldots\ldots\ldots\ldots & g_N((K-2)L_s) \\ \vdots & \vdots & \vdots \\ g_N(-(K-1)L_s)\ldots\ldots & g_N(-(K-2)L_s)\ldots\ldots\ldots & g_N(0) \end{bmatrix} \tag{6.11}$$

An eigenvalue decomposition (EVD) is then performed on matrix $\overline{\mathrm{A}}$. These resultant eigenvalues are later used for minimization of the variance of the estimated result.

Thus, the estimated result from FBSE can be written as the time average of the square magnitude of the output of the mth sub-band filter. Mathematically,

$$\hat{\Gamma}_{FBSE}(f_m) = \frac{1}{K}\sum_{k=0}^{K-1} |y_m[n - kL_s]|^2 \tag{6.12}$$

Advantages:

- As a root-Nyquist (only one prototype filter) filter is used in FBSE, which has the best magnitude response of the prototype filters in MTSE, it performs better in terms of variance.
- For the better magnitude response of the prototype filter, FBSE introduces lower leakages.
- For a large number of samples FBSE performs better than MTSE.

Disadvantages: MTSE is better in accuracy and speed of estimation issues for small number of samples because of smaller window size.

From the above discussions it is seen that conventional spectrum estimation techniques suffer from several drawbacks. Though leakages and reduction of variance of the estimated signal are controlled by MTSE and FBSE, they still suffer from optimal design of the time–frequency resolution window. For cognitive radio, if the spectrum under test is highly non-stationary, then the above methods face severe design challenges.

Recent studies indicate implementation of the wavelets and wavelet transform for the detection of singularities and edges of the spectrum, as is proposed by [10]. A wavelet transform has the greatest advantage to dynamically adjust time and frequency resolution, in accordance with a highly non-stationary signal. This property can be implemented in the estimation of the highly varying radio spectrum in cognitive radio. In [11] the idea of a spectrum estimator based on wavelet packets (a modified version of wavelet transform) and filter bank is proposed.

6.5 Advantages of wavelets in spectrum estimation

Based on the survey of all the traditional and modern spectrum sensing methodologies [12], it is seen that a low complexity spectrum estimator with the property of a flexible time–frequency resolution is desirable. Wavelets have the property of dynamic tunability of time–frequency resolution but there is no fixed rule to select wavelets for any specific operation. Some advantages of wavelets for spectrum sensing operation can be summarized as follows.

1. One of the most important advantages of wavelets lies in their ability to tune the time–frequency window (maintaining the orthogonality) in a manner that can detect the dynamic variation of the statistical parameters of any spectrum. The reconfigurable structure of the transform introduces adaptability of the system based on channel, link and other transmission characteristics.

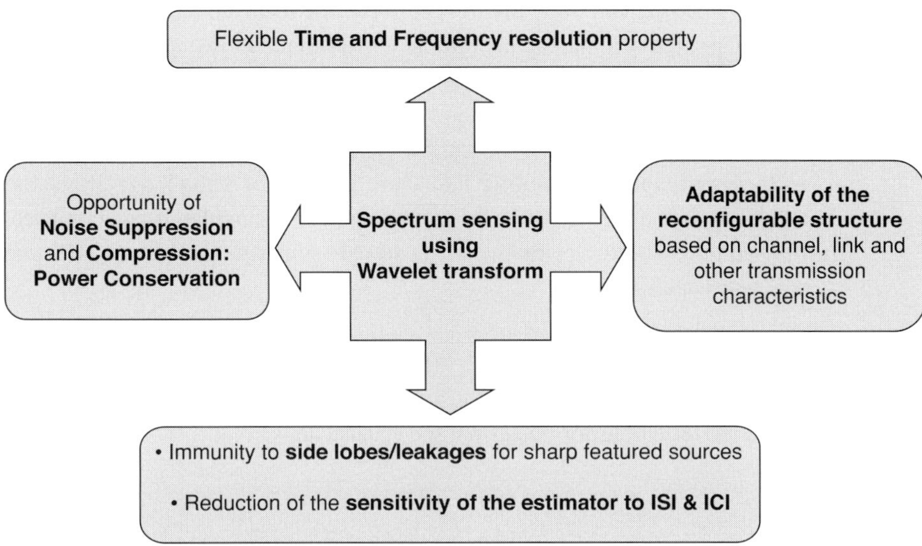

Figure 6.5 Advantages of spectrum sensing using wavelets.

2. By flexible design of the time–frequency window, the effect of noise and interference on the signal can be minimized. The wavelet-based de-noising is common in the case of many image signals.
3. Wavelet based algorithms can be implemented for significant data compression (already implemented in digital image compression). By compressing the data, the total number of sensing measurements may be reduced. This phenomenon reduces the communication power needed for transmission and simultaneously the speed of the estimation is also enhanced.
4. The tuning flexibility of wavelet bases has advantages like mitigation of channel effects such as inter-symbol interference (ISI) and inter-carrier interference (ICI).
5. Recent results [11] also show that a wavelet-based estimate has fewer side-lobes and leakages than some traditional methods for sharp-featured sources.

These advantages are illustrated in Figure 6.5.

6.6 Performance evaluation of spectrum sensing in cognitive radio

6.6.1 Basic principle of energy detector

The detection of any signal in an additive white Gaussian noise environment is analyzed by the well-known binary hypothesis test problem based on the Neyman–Pearson criterion [13].

This can be formulated by,

$$\left.\begin{array}{lll} r[n] = w[n], & n = 1, 2, \ldots, N & : H_0 \\ r[n] = x[n] + w[n], & n = 1, 2, \ldots, N & : H_1 \end{array}\right\} \tag{6.13}$$

where $r[n]$, $x[n]$ and $w[n]$ are, respectively, the received signal at cognitive user nodes, the transmitted original signal and the noise signal whose samples are from a zero mean Gaussian random process with constant power spectral density σ_n^2. Hypotheses H_0 and H_1 correspond to the absence and presence of the target signal (primary or licensed user in the case of cognitive radio), respectively. N is the total number of samples of the input signal. Assuming that there is no prior knowledge about the transmitted signal, in the energy detector the average power is calculated and compared with a predefined threshold to decide the target signal's presence or absence. The energy detector is constructed as [14],

$$X(r) = \frac{1}{N}\sum_{n=1}^{N} r^2[n] \tag{6.14}$$

If λ is the threshold for detection, the criteria are,

$$\begin{aligned} X[r] &> \lambda \quad : H_1 \\ X[r] &< \lambda \quad : H_0 \end{aligned} \tag{6.15}$$

The probability of detection and false alarm is defined in the following manner.

Probability of detection: The conditional probability of detecting the target signal to be present subject to the condition that actually it is present, known as the probability of correct detection (P_d).

Probability of false alarm: The conditional probability of detecting the target signal to be present subject to the condition that actually it is absent, known as the probability of false alarm (P_{fa}).

In the cognitive radio paradigm the target can be a primary user or a spectrum hole.

If S_{av} is the average signal power and σ_n^2 is the noise variance, then the false alarm and detection probabilities can be formulated as [14],

$$P_{fa} = Q\left(\frac{\lambda - \sigma_n^2}{\sqrt{\frac{2\sigma_n^4}{N}}}\right) \tag{6.16}$$

$$P_d = Q\left(\frac{\lambda - (S_{av} + \sigma_n^2)}{\sqrt{\frac{2(S_{av}+\sigma_n^2)^2}{N}}}\right) \tag{6.17}$$

where, $Q(x) = \frac{1}{\sqrt{2\pi}}\int_x^\infty e^{\frac{-u^2}{2}}du$ is the Gaussian complementary cumulative distribution function. Expressions (6.16) and (6.17) are the relations that exhibit the theoretical false alarm and detection probabilities in terms of signal and noise power.

6.6.2 Evaluation of receiver operating characteristics (ROC)

The variation of the probability of detection with the false alarm probability for a wide range of thresholds gives the theoretical receiver operating characteristics (ROC). For

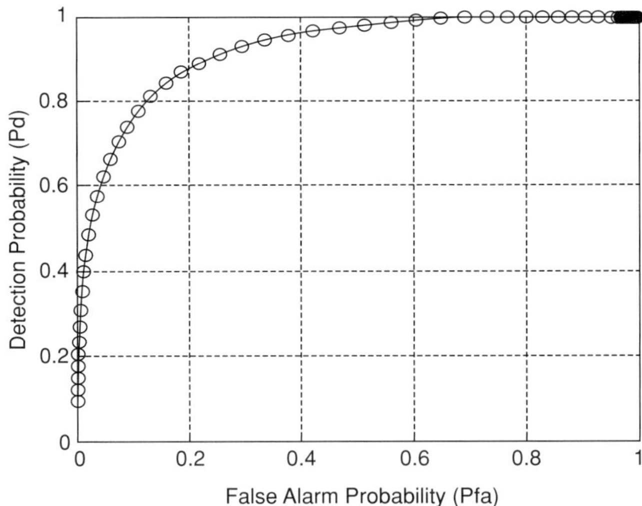

Figure 6.6 Receiver operating characteristics (theoretical).

any CR spectrum sensing methodology, the ROC curve is one of the most important performance metrics. A high detection probability with a very low false alarm is always desirable. Figure 6.6 exhibits the nature of ROC.

6.7 Wavelet packet spectrum estimator (WPSE)

Recently, the wavelet-packet-based spectrum estimator (WPSE) was proven to be an efficient method for spectrum estimation for cognitive radio applications [19]. From the mathematical theory of wavelets it is seen that wavelets of compact support can be realized from perfect reconstruction of filter banks [15]. In [11] an approach for the application of the wavelet packet transform for spectrum sensing in cognitive radio is illustrated. The major advantage of the wavelet packet transform for spectrum estimation is the orthogonality of the filters that makes the filtered output uncorrelated, and consequently minimizing the bias in the estimate.

Wavelet packet transforms (WPT) successively split the input signal into high- and low-frequency components. The transform produces a binary tree-based decomposition of the input signal. So, for an L level of decomposition the transform will generate a binary tree with 2^L terminal nodes as the leaves of the complete tree. Each terminal node contains a set of wavelet-packet coefficients of finite length. As depicted in Figure 6.7 the two channel filter banks tree structure recursively decomposes the signal being estimated and maps the signal components into the different frequency bands. The output of each wavelet packet node corresponds to a particular frequency band. An L level wavelet packet decomposition divides the frequency band ranging from 0 to 0.5 f_s (where f_s denotes the sampling frequency) in 2^L frequency sub-bands [15]. The decomposition of the signal into different frequency bands with different resolutions is possible. Due to

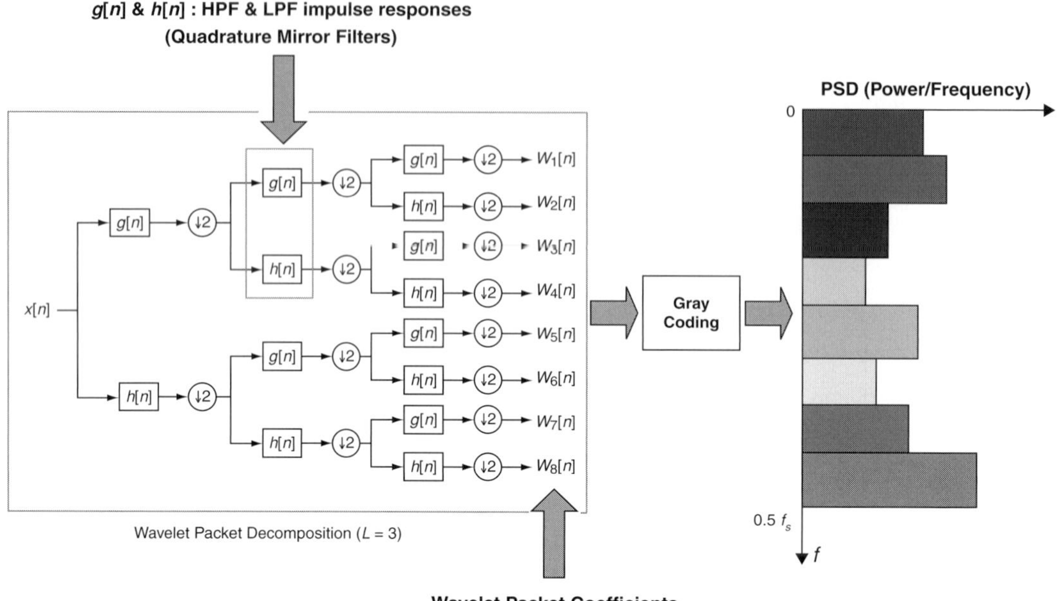

Figure 6.7 Representation of a 3-level WPSE; $g[n]$ and $h[n]$ are, respectively, the impulse responses of high-pass and low-pass quadrature mirror filters. $w_1[n], w_2[n], w_3[n], \ldots, w_5[n]$ are the wavelet packet coefficients.

filtering and down-sampling of the input signal, at every level in wavelet-packet based decomposition time resolution is halved and the frequency resolution is doubled. Thus, the frequency resolution of the estimation is related to the level of decomposition.

The most important issue of this scheme is that the impulse responses of the high-pass ($g[n]$) and low-pass filters ($h[n]$) must be coupled by a quadrature mirror filters (QMF) relation ($g[n] = (-1)^n h[L - 1 - n]$, for $h[n]$ with length of filter L) to establish the orthogonality between scaling and wavelet functions.

One of the most interesting features of WP-based spectrum estimation is that the coefficients of the wavelet packet transform are not naturally ordered by increasing order of frequency. They are rather numbered on the basis of a sequential binary grey code value [16]. For example, if each coefficient in the level basis is numbered with a sequential decimal order (0000, 0001, 0010, 0011, . . .), the frequency ordering of the coefficients can be ordered by frequency by sorting them into grey code values (0000, 0001, 0011, 0010, 0110, . . .).

From the mathematical analysis shown in [11], it is seen that the wavelet packet transform can be successfully implemented for the spectrum sensing as it preserves the time-domain energy in the wavelet domain satisfying Parseval's relationship.

An L level wavelet packet decomposition divides the frequency band ranging from 0 to $0.5\, f_s$ (where f_s denotes the sampling frequency) into 2^L frequency sub-bands [15]. The energy contained in a certain band (Joule) can be found from the inner product of wavelet coefficients vector of the corresponding node with itself. The power spectral

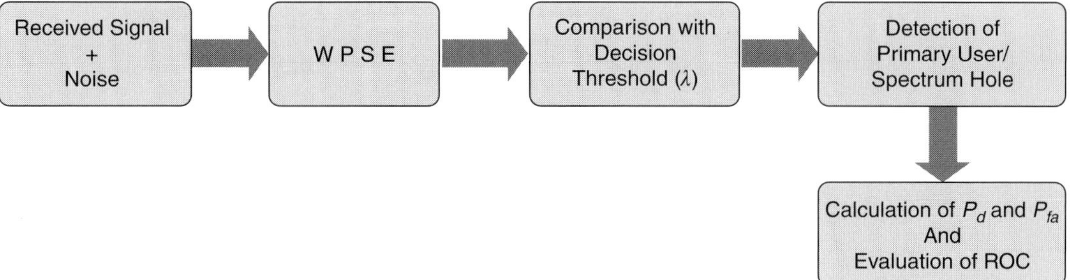

Figure 6.8 Method of evaluation of ROC in WPSE.

density (PSD) at the kth frequency band can be written as,

$$\mathrm{PSD}(f_k) = \frac{E(f_k)}{N_s f_k} \tag{6.18}$$

where N_s is the total number of samples in each wavelet-packet coefficients at the Lth decomposition level, f_k is the frequency range spanned by kth wavelet packet node and $E(f_k)$ is the energy (inner product of the wavelet coefficient vector of the corresponding node with itself) of the kth wavelet packet node.

6.7.1 Evaluation of ROC performance of WPSE

The method of evaluation of ROC in WPSE is discussed in [17]. In this section ROC is evaluated in WPSE and the nature of ROC is studied for different types of wavelet families. The variation of the ROC in WPSE is also studied for different wavelet filter properties by simulation.

The technique of studying the ROC is illustrated in Figure 6.8.

1. The input signal corrupted with noise is decomposed by an L-level WPSE. The signal is divided in 2^L sub-bands in the frequency range of 0 to $0.5f_s$ (where f_s is the sampling frequency of the input signal).
2. Powers in individual sub-bands are estimated by the mathematical formulation of Eq. (6.18) and each sub-band power is compared with a pre-defined decision threshold (λ), and a decision is made if the sub-band is a primary/active user or a spectrum hole.
3. Detection and false alarm probabilities are calculated from the total number of correctly or wrongly detected active/primary users and the actual number of active/primary users, respectively. Receiver operating characteristics (ROC) are evaluated by plotting P_d versus P_{fa}.

6.7.1.1 Simulation results

The ROC of WPSE is evaluated for different wavelet families (both orthogonal and non-orthogonal) and wavelet filter parameters like length of the filter, number of primary users and signal-to-noise ratio. To evaluate the ROC of WPSE, we have taken N_c single

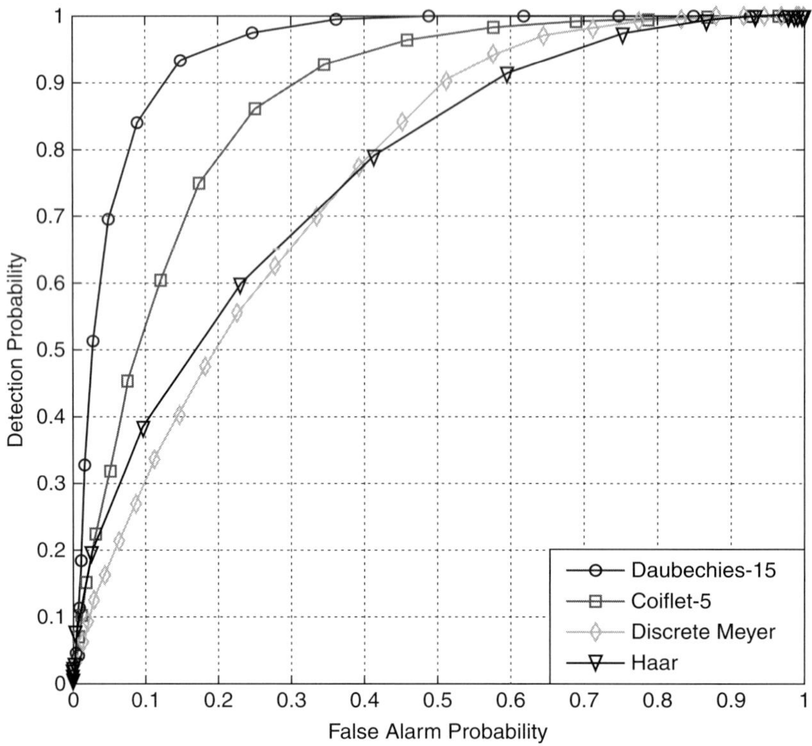

Figure 6.9 Variation of ROC for different wavelet families; SNR = 0 dB; level of decomposition (L) = 7; number of single tones as PU (2^L) = 128; threshold for detection (λ) = –20 to 0 dB.

tones as primary users in cognitive radio that constitutes the entire normalized frequency band of $[0,\pi]$ (radian/sample). The sampling frequency is taken as 2π radian/sample. For simplicity of the analysis we assumed that $N_c = 2^L$ for the L level of decomposition. To estimate the probability of false alarm and probability of correct detection, all N_c single tones/carriers are randomly activated and deactivated.

This carrier activation and deactivation process is the same as carrier activation and deactivation for orthogonal frequency division multiplexing (OFDM) [18] signals. ROC in WPSE is evaluated using different wavelet families and their performance is analyzed. The detection and false alarm probabilities are calculated out of 100 iterations. For every iteration, the detection threshold (λ) for active/primary user is varied in the range of –20 to 0 dB. For simplicity the channel is assumed to be an additive white Gaussian noise (AWGN) channel.

Figure 6.9 shows the variation of ROC in WPSE for different standard wavelet families. It is seen that the Daubechies wavelet has the best performance because of its better frequency selectivity than the others. The performance of the Haar wavelet is poor because of its small length of support. In the above analysis all the wavelets used were orthogonal.

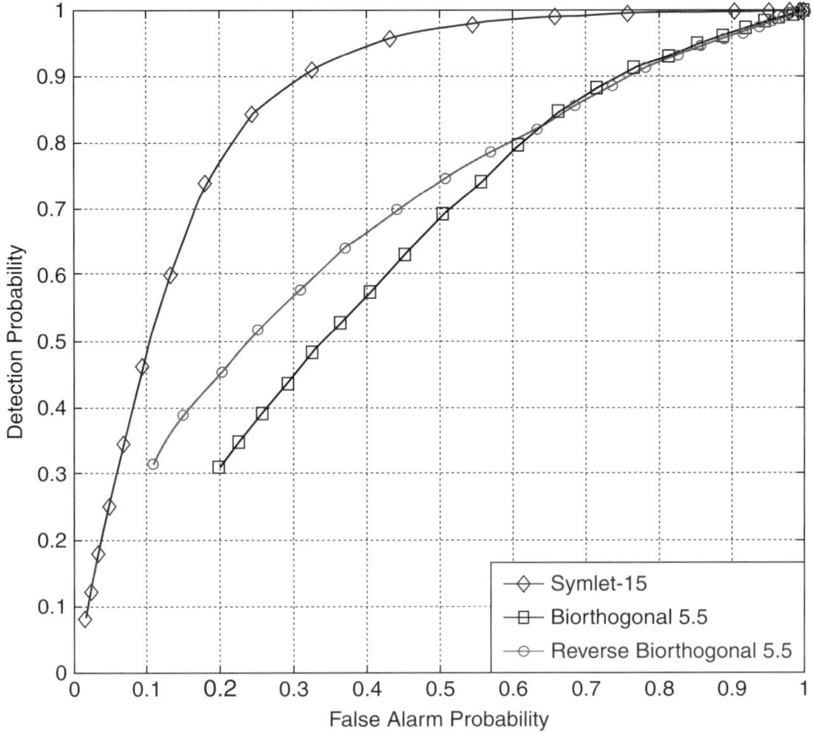

Figure 6.10 Variation of ROC for different wavelet families (both orthogonal and non-orthogonal); SNR = 0 dB; level of decomposition (L) = 7; number of single tones as PU (2^L) = 128; threshold for detection (λ) = –20 to 0 dB.

In the next case both orthogonal and non-orthogonal wavelets are used and the result is shown in Figure 6.10.

It is clear that as biorthogonal and reverse biorthogonal wavelets are not orthogonal, the performance of ROC with a Symlet 15 wavelet is much better than biorthogonal and reverse biorthogonal wavelets. This analysis shows that orthogonality of any wavelet filter is one of the main criteria for achieving a good frequency selectivity for WPSE.

In Figure 6.11 the variation of ROC performance is shown with the length of wavelet filters.

From Figure 6.11 it is seen that as the length of wavelet filters increases the performance gets better. The behaviour of the curves in Figure 6.11 is due to the reduction of variance of the estimated PSD in the unoccupied band and the improvement of frequency selectivity with increasing length of wavelet filters.

In Figure 6.12 the variation of ROC with the number of primary users is shown. The total band of $[0,\pi]$, used for simulation consists of N_c single tones as primary users, which are randomly activated or deactivated. As the level of decomposition (L) increases the number of single tones or primary users is also increased, as we have assumed

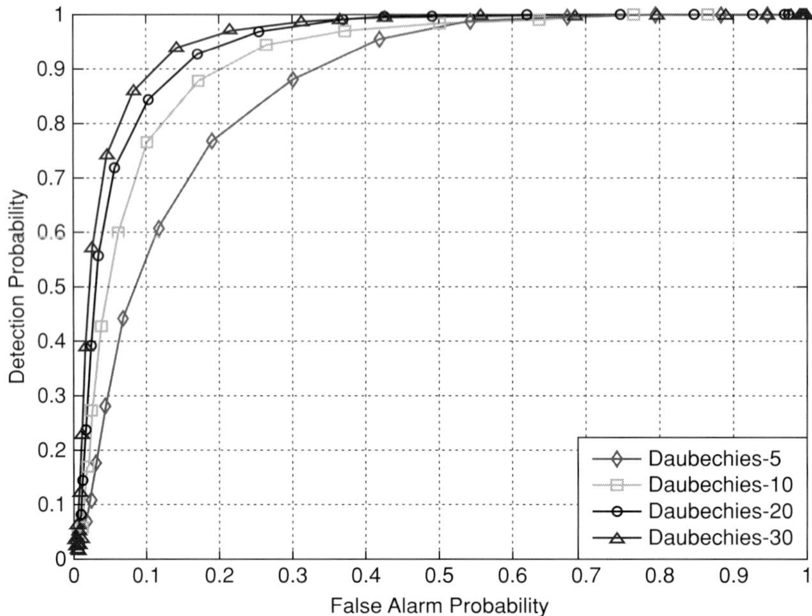

Figure 6.11 Variation of ROC with different length of wavelet filters; SNR = 0 dB; level of decomposition (L) = 7; number of single tones as PU (2^L) = 128; threshold for detection (λ) = –20 to 0 dB.

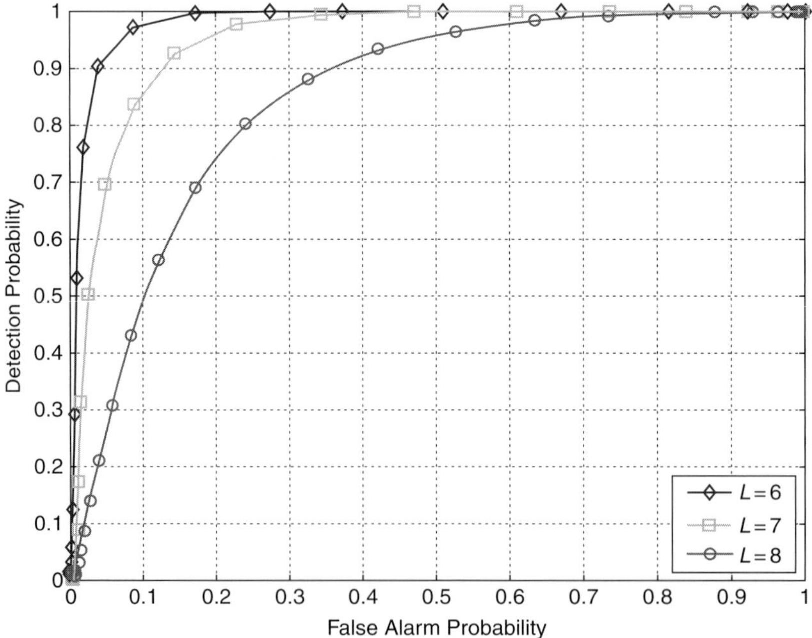

Figure 6.12 Variation of ROC with different numbers of primary users; SNR = 0 dB; level of decomposition (L) = 7; number of single tones as PU = 2^L; threshold for detection (λ) = −20 to 0 dB.

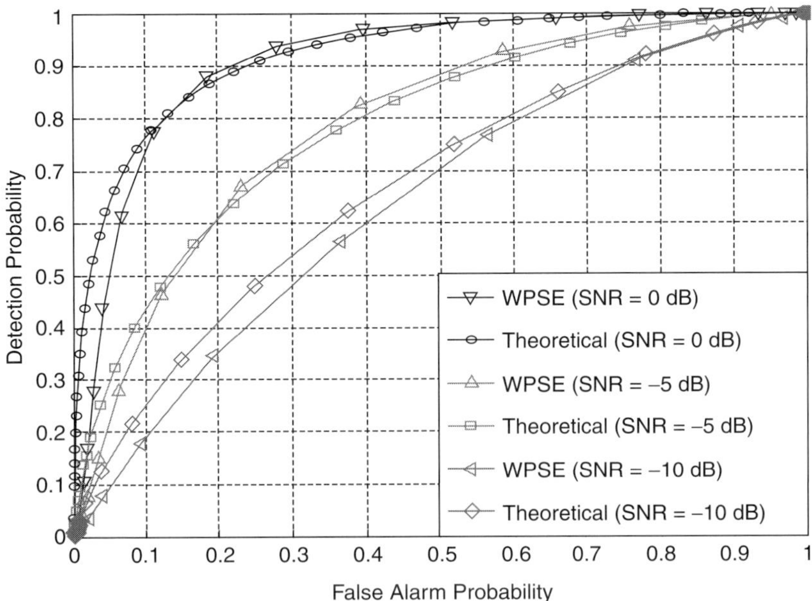

Figure 6.13 Variation of theoretical ROC and ROC of WPSE (simulated) with different SNR; level of decomposition $(L) = 7$; number of single tones as PU $= 2^L$; threshold for detection $(\lambda) = -20$ to 0 dB.

$N_c = 2^L$. So, at the high level of decomposition, the frequency selectivity for wavelet-packet-based estimation of any single tone increases but simultaneously the amount of side-lobe power for the estimation of individual single tone is also increased. Thus, in the case of higher numbers of primary users the overall false alarm rate for the entire band increases.

In the previous section the signal-to-noise ratio (SNR) was kept constant for all the analyses. The theoretical variation of probability of correct detection and probability of false alarm with SNR can be achieved from Eqs. (6.16) and (6.17). In this section, the variation of the detection and false alarm probabilities for WPSE is studied with different SNR and the results are compared with theoretical values. Figure 6.13 shows the ROC for both theoretical and WPSE. The noise added is assumed to be additive white Gaussian (AWGN). In the case of WPSE it is seen that detection performance improves with increasing SNR, which is analogous to the theoretical results of Eqs. (6.16) and (6.17).

6.8 An efficient model of wavelet-packet-based spectrum estimator

6.8.1 WPSE model

In [19], an efficient model of WPSE with reduced number of sensing measurement is proposed. The reduction of sensing measurements is a type of compression induced

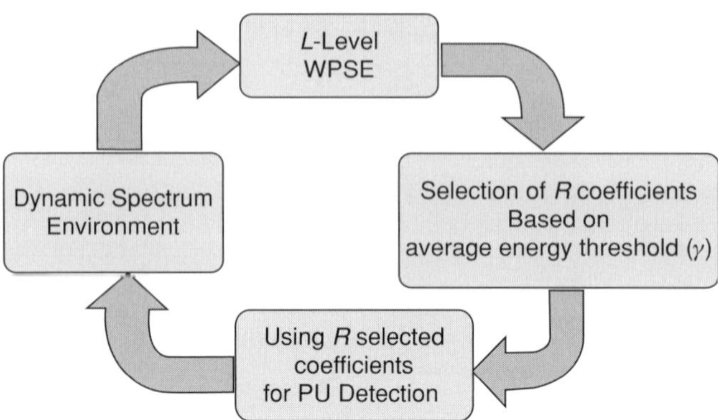

Figure 6.14 Block diagram of the developed model for efficient detection of primary user (PU).

in the sensing and detection of the primary user. From the mathematical theory of compressed sensing [20], [21] we can see that it is possible to reconstruct a signal with N samples from M random linear measurements where, $M \ll N$. Then, the signal x is said to be compressible subject to the condition that it must be sufficiently sparse [22] in any particular basis; i.e., a very small portion of all the signal coefficients carries the essential information (very few non-zero elements in the signal vector). In the cognitive radio paradigm the radio spectrum also exhibits temporal and spatial sparse behaviour [4], [5]. However, utilizing the sparsity of the wideband spectrum, a method for spectrum estimation and detection from sub-Nyquist samples with a wavelet-based edge detector is already proposed in [23].

In Figure 6.14 the model for detection of the primary user with a reduced number of sensing measurements is illustrated. In this scheme first the dynamic environment is sensed by a wavelet-packet-based spectrum estimator. It is seen that for an L level of decomposition there will be 2^L terminal coefficients. In the next step only R coefficients are selected out of 2^L. This selection of R coefficients is done based on the average energy content of the signal. The threshold (γ) for wavelet-packet coefficient selection can be set dynamically based on the spectrum environment or any pre-defined statistical record regarding the power spectral density (PSD) of the primary user. As it is already seen that a significant part of the spectrum is underutilized for certain times or geographic locations, the number of coefficients with high energy content will be much smaller than the total number of coefficients. So, these few coefficients, R ($R < 2^L$) can be used for the detection of the location of active users and informing the secondary/cognitive users for opportunistic spectrum access. The advantage of the above-mentioned scheme is the reduced number of sensing measurements. Only $M = R/\, 2^L$ of the total coefficients are used to detect the spectral location of active/primary user for any level of decomposition L.

It is clear that if a small portion of the spectrum is occupied by primary user, which is the case for many real situations, R and consequently M will be small, reducing the total sensing load for primary user detection. Selection of the threshold (γ) for wavelet-packet coefficient selection is an important issue in this scheme. Here, we have selected

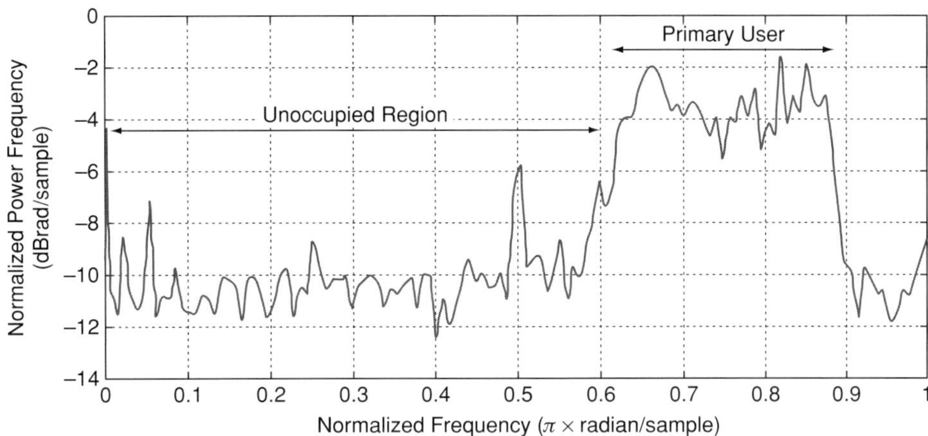

Figure 6.15 Wavelet-packet-based estimate of spectrum $[0,\pi]$, whose 30% is occupied by primary user; SNR = 0 dB; wavelet = Daubechies 20.

the threshold on the basis of the estimated (WPSE) active user power derived from the wavelet packet transform of the signal with fixed SNR.

For this stage the situation is considered to be static. However, the information (coefficients indicating the presence of primary user) retrieved at one instant of time can be used for another instant for slowly varying dynamic environments.

6.8.2 Study of the detection performance of the developed model

To study the detection performance of the developed model, only a small portion of the entire spectrum with normalized frequency of $0 - \pi$ (radian/sample) that is occupied by the primary user is considered (Figure 6.15). The approximate width (0.3π) of the primary user (partial sub-band) is fixed arbitrarily. The width of the partial sub-band is kept intentionally low (30% of the entire spectrum) keeping the significant underutilization of the radio spectrum in mind.

Keeping the width of the occupied portion of the spectrum constant in the simulation, the location of this occupied partial sub-band is varied randomly throughout the entire spectrum and detection and false alarm probabilities are calculated out of 1000 iterations of primary user spectral locations. The decision threshold for detection is kept constant at –10 dB. The threshold for wavelet-packet coefficient selection (γ) is varied from –30 to –3 dB resulting in different compression rates. Figure 6.16 exhibits the variation of detection performance with the compression rate ($R/2^L$) with Daubechies 15 and Symlet 20 wavelet families.

From Figure 6.16 it is seen that only 25–30% of the total coefficients can achieve 100% probability of detection for a Daubechies 15 wavelet. The false alarm probability (P_{fa}) is almost constant at 0.01–0.02, which is analogous to theoretical results because of the fixed value of the decision threshold (λ) in Eq. (6.16). The slight increase of false alarm probability near the compression rate of 0.25 is due to the side-lobes of

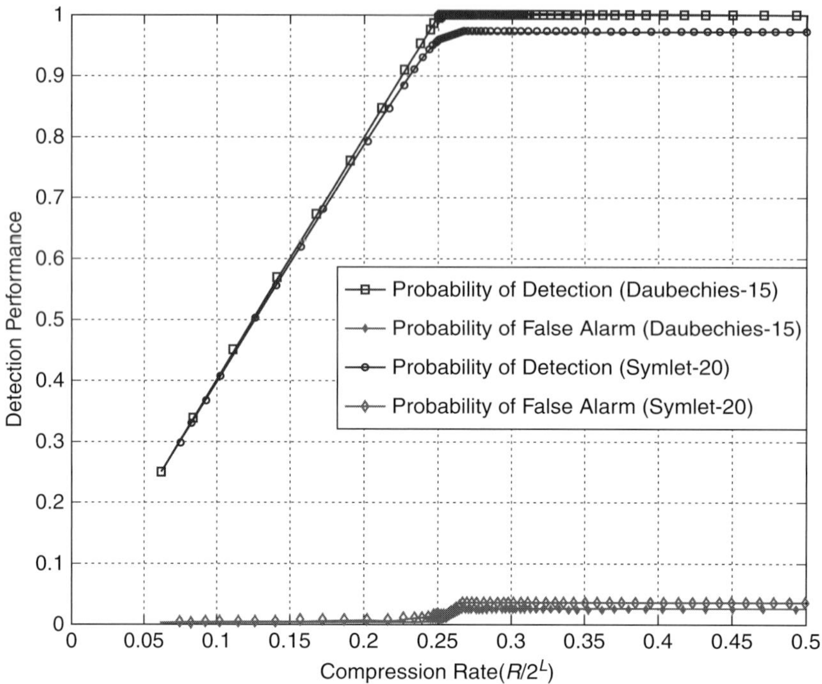

Figure 6.16 Variation of detection performance with compression rate; SNR = 0 dB.

the wavelet-packet-based estimator, particularly at lower threshold levels for coefficient selection (γ).

6.9 Wavelet-packet-based spectrum estimator (WPSE) and compressed sensing

6.9.1 Introduction to compressed sensing

Compressed sensing [20], [21] is a newly developed mathematical tool that states that it is possible to reconstruct a signal with N samples from M random linear measurements where, $M \ll N$. Then, the signal x is said to be compressible subject to the condition that it must be sufficiently sparse [22] in any particular basis; i.e., a very small portion of all the signal coefficients carries the essential information (very few non-zero elements in the coefficient vector).

Any signal in R^{N} can be represented in terms of a basis of $N \times 1$ vectors $\{\psi_i\}_{i=1}^{N}$. For simplicity, it is assumed that the basis is orthonormal. Forming the $N \times N$ basis matrix by stacking the vectors $\{\psi_i\}_{i=1}^{N}$ as columns, any signal $x\,(t)$ can be expressed as

$$x(t) = \sum_{i=1}^{N} s_i \psi_i(t) \text{ or } x = \Psi s \tag{6.19}$$

where s_i is the $N \times 1$ column vector of weighting coefficients are given by $s_i = < x, \psi_i >$. The Ψ matrix is known as the *representation matrix*. Clearly, x and s are equivalent representations of the same signal, with x in the time domain and s in the Ψ domain. The signal x is *K-sparse* if only K of the all N coefficients are non-zero ($K \ll N$). The rest ($N - K$) coefficients are all zeros.

In the compressed sensing paradigm it is possible to directly recover the signal by calculating $M < N$ inner products between x and a collection of vectors $\{\varphi_j\}_{j=1}^{M}$, as in $y_j = < x, \varphi_j >$.

Stacking the measurements y_j into $M \times 1$ vector y, and the measurement vectors $\{\varphi_j\}^T$, as rows into an $M \times N$ matrix Φ and substituting in Eq. (6.19),

$$y = \Phi x = \Phi \Psi s = \Theta s \tag{6.20}$$

Where $\Theta = \Phi \Psi$ is an $M \times N$ matrix and Φ is also an $M \times N$ matrix (also known as *the measurement matrix*). Reconstruction of the signal from Eq. (6.20) leads to an optimization problem.

The three main conditions to implement compressed sensing are 1. Sparsity, 2. Incoherence, 3. Restricted isometry property (RIP). The details of these mathematical conditions are illustrated in [21], [20]. The main objective is to design the measurement matrix and a suitable reconstruction algorithm for the *K-sparse* and compressible signals that require only M ($K < M \ll N$) measurements for exact recovery. In [24] an example of compressed sensing of an original image and the wavelet transform of the image, as well as a successful reconstructed image by discarding 97.5% of the coefficients, is reported.

6.9.2 Compressed sensing and WPSE

Sparsity is one of the most important criteria for compressed sensing. In the wavelet-packet domain it is seen that in some signals, after a high level of wavelet-packet decomposition, the wavelet-packet coefficients also exhibit sparse behaviour. In Figure 6.17 and Figure 6.18 it is seen that the wavelet-packet coefficients for the signal that has two and three primary users as partial sub-bands also exhibit sparse behaviour. It is also seen that as the number of primary users is increased the amount of sparsity reduces (the value of K increases, as discussed in the previous section).

Establishment of WPSE in the light of the mathematical theory of compressed sensing is the state-of-the-art. It involves the study of the following features:

- Analysis of the sparsity of the signal in the wavelet domain or in any other basis.
- Selection/design of the proper basis for compressed sensing that satisfies the criteria for compressed sensing.
- Analysis of the sparse tree-based representation of signal.

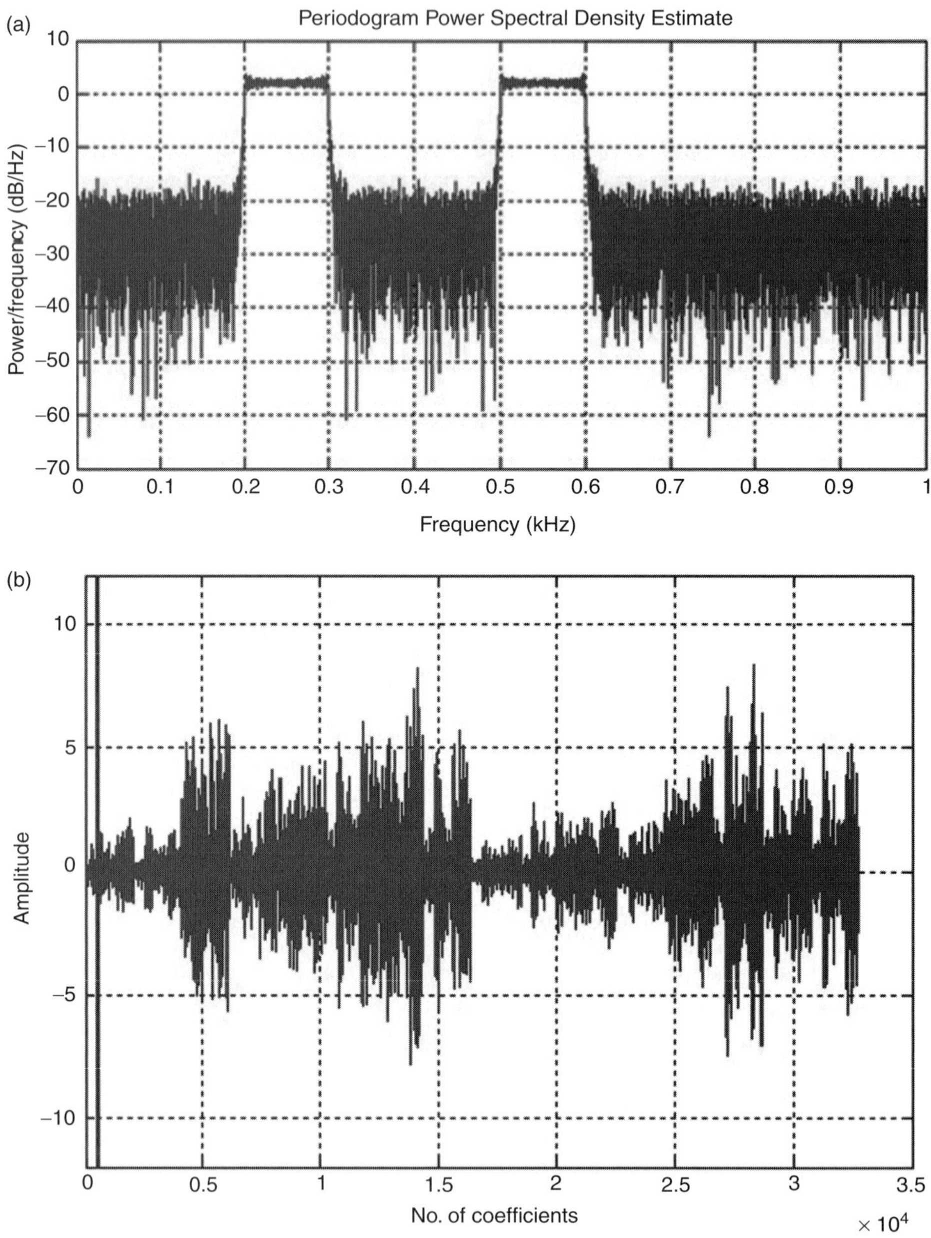

Figure 6.17 (a) Fourier-based estimate of a signal having 2 primary users (SNR = 10 dB). (b) Sparse representation of the input signal; wavelet = Haar; $L = 15$.

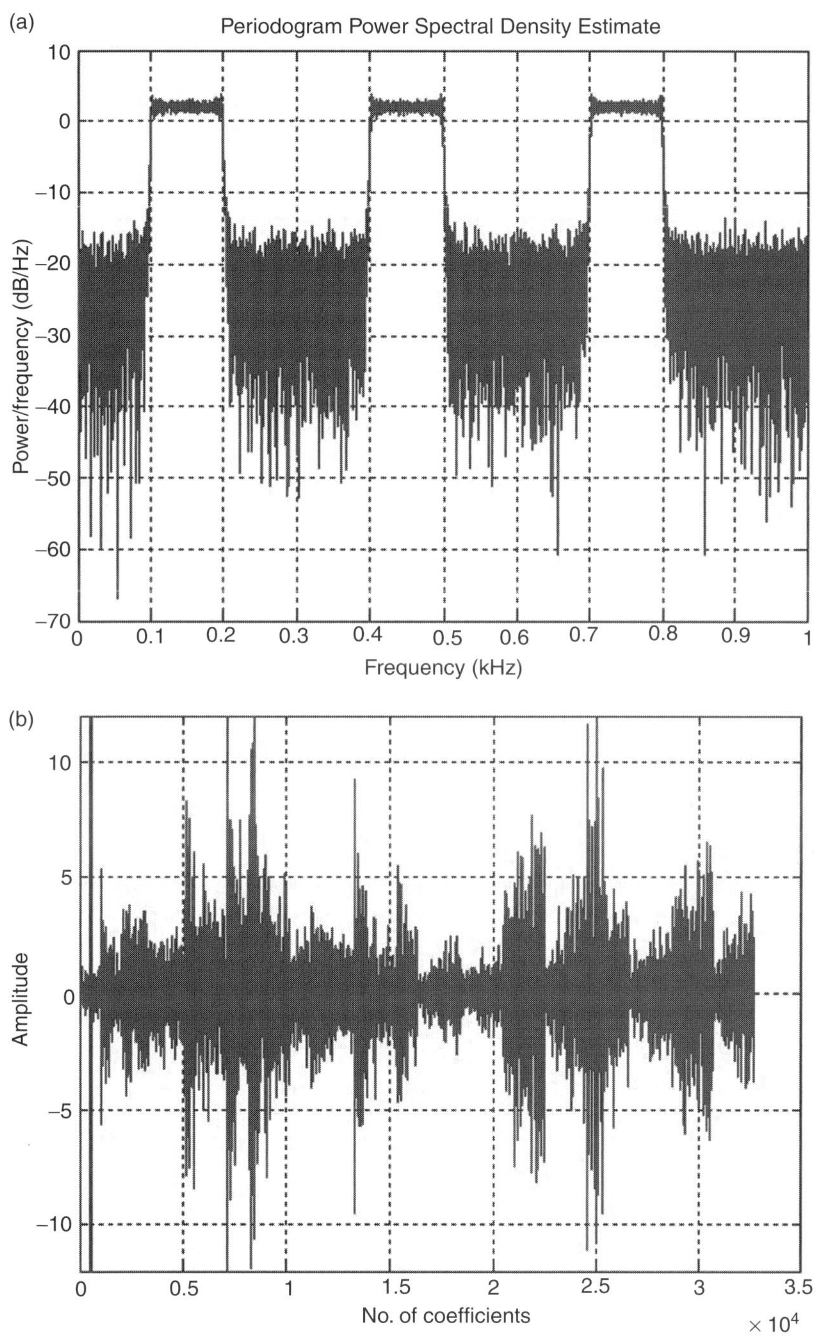

Figure 6.18 (a) Fourier-based estimate of a signal having 3 primary users (SNR = 10 dB). (b) Sparse representation of the input signal; wavelet = Haar; $L = 15$.

The possible achievements from the above analysis can be stated as:

- Reduction of number of sensing measurements.
- Reduction of complexity of the spectrum sensing method.
- Optimized use of sensing power.
- Optimized number of sensing nodes.

In summary, it can be stated that the compressed sensing technique along with the wavelet packet spectral estimation proposes an efficient and sustainable (green) solution for cognitive radio spectrum sensing.

6.10 Summary

In this chapter, we studied the variation of the receiver operating characteristics (ROC) for a wavelet-packet-based spectrum estimator (WPSE) with different types of wavelet families and wavelet filters. It was observed that the orthogonality of the wavelet filters is one of the major criteria for achieving a good detection performance. Secondly, as the length of the wavelet filters is increased, the frequency selectivity is also improved, giving rise to better detection performances. However, as the number of primary users is increased, the performance deteriorates due to more side-lobes in the estimation of individual primary users. The improvement of detection performance with increasing signal-to-noise ratio of the input signal was studied for both WPSE (simulated) and the theoretical ROC. An efficient model of WPSE, with a reduced number of sensing measurements, was also developed. Finally, it was exhibited that the selection of a wavelet-packet coefficient based on average energy threshold reduces the number of sensing measurements for the detection of the primary user. Results showed that, using the developed model, for 30% occupancy of the spectrum, only 25–30% of all the wavelet-packet coefficients can achieve almost 100% probability of detection.

The future research is envisioned towards the development of an efficient model of WPSE in the light of the mathematical theory of compressed sensing utilizing the temporal and spatial sparse behaviour of the radio spectrum. This study also includes proper selection/design of basis to implement compressed sensing in WPSE. Study of the dynamic behaviour and the detection performance of the developed model in different channel conditions and propagation environments also indicates the future scope of the work.

References

[1] J. Mitola III, "Cognitive Radio: An Integrated Agent Architecture for Software Defined Radio," Doctoral Dissertation, Royal Institute of Technology (KTH), Sweden, May 2000.

[2] S. Haykin, "Cognitive Radio: Brain-Empowered Wireless Communications," *IEEE Journal on Selected Areas in Communications*, vol. 23, No. 2, pp. 201–20, February 2005.

[3] G. Staple and K. Werbach, "The End of Spectrum Scarcity," *IEEE Spectrum Archive*, vol. 41, no. 3, pp. 48–52, March 2004.

[4] Federal Communication Commission, "Spectrum Policy Task Force Report," ET Docket No. 02–135, November 2002.

[5] Federal Communication Commission-First Report, and Order and Further Notice of Proposed Rulemaking, "Unlicensed operation in the TV broadcast bands" FCC 06–156, Oct. 2006.

[6] D. Cabric, S.M. Mishra, and R.W. Brodersen, "Implementation Issues in Spectrum Sensing for Cognitive Radios," *Proc. of the Asilomar Conference on Signals and Systems*, 2004.

[7] D. Cabric, A. Tkachenko, and R.W. Brodersen, "Spectrum Sensing Measurements of Pilot, Energy, and Collaborative Detection," *IEEE Military Communications Conference*, October 2006.

[8] D.J. Thomson, "Spectrum Estimation and Harmonic Analysis," *Proc. IEEE*, vol. 70, no. 9, pp. 1055–96, September 1982.

[9] B. Farhang-Boroujeny, "Filter Banks Spectrum Sensing for Cognitive Radios," *IEEE Transactions on Signal Processing*, vol. 56, pp. 1801–11, May 2008.

[10] Z. Tian and G.B. Giannakis, "A Wavelet Approach to Wideband Spectrum Sensing for Cognitive Radios," *Proceedings of International Conference on Cognitive Radio Oriented Wireless Networks and Communications*, Greece, 2006.

[11] D.D. Ariananda, M.K. Lakshmanan, and H. Nikookar, "A Study on Application of Wavelets and Filter Banks for Cognitive Radio Spectrum Estimation," *European Wireless Technology Conference (EuWiT2009)*, Rome, Italy, October 2009.

[12] D.D. Ariananda, M.K. Lakshmanan, and H. Nikookar, "A Survey on Spectrum Sensing Techniques for Cognitive Radio," *2nd International Workshop on Cognitive Radio and Advanced Spectrum Management, CogART*, 2009.

[13] H.L. Van Trees, *Detection, Estimation and Modulation Theory, Part 1*, John Wiley and Sons, 2001.

[14] R. Tandra and A. Sahai, "SNR Walls for Signal Detection," IEEE *Journal of Selected Topics in Signal Processing*, vol. 2, No. 1, pp. 4–17, Feb. 2008.

[15] M. Vetterli and I. Kovacevic, *Wavelets and Subband Coding*, Englewood Cliffs, New Jersey, Prentice-Hall PTR, 1995.

[16] A. Jensen and A. la Cour-Harbo, *Ripples in Mathematics: The Discrete Wavelet Transform*, Springer, Germany, 2001.

[17] D.D. Ariananda, M.K. Lakshmanan, and H. Nikookar, "Design of Best Wavelet Packet Bases for Spectrum Estimation," *Proc. of the 20th Personal, Indoor and Mobile Radio Communications Symposium 2009 (PIMRC'09)*, September 2009.

[18] T. Weiss and F.K. Jondral, "Spectrum Pooling: An Innovative Strategy for the Enhancement of Spectrum Efficiency," *IEEE Communications Magazine*, 42, S8–S14, March 2004.

[19] V. Roy and H. Nikookar, "Performance Evaluation of a Wavelet Packet-based Spectrum Estimator for Cognitive Radio Applications," *18th IEEE Symposium on Communications and Vehicular Technology in the Benelux*, Ghent, Belgium, 22–23 November 2011.

[20] D.L. Donoho, "Compressed Sensing," *IEEE Transactions on Information Theory*, vol. 52, No. 4, pp. 1289–306, Apr. 2006.

[21] E. Candes, J. Romberg, and T. Tao, "Robust Uncertainty Principle: Exact signal reconstruction from highly incomplete frequency information," *IEEE Transactions on Information Theory*, vol. 52, no. 2, pp. 489–509, Feb. 2006.

[22] E. Candes and J. Romberg, "Sparsity and Incoherence in Compressive Sampling," *Inverse Problems*, 23(3), pp. 969–85, June 2007.

[23] Z. Tian and G.B. Giannakis, "Compressed Sensing for Wide-band Cognitive Radios," *Proc. of International Conference on Acoustics, Speech and Signal Processing*, pp. IV/1357–IV/1360, April 2007.

[24] E.J. Candes and M.B. Wakin, "An Introduction to Compressive Sampling," *IEEE Signal Processing Magazine*, vol. 25, no. 2, pp. 21–30, 2008.

[25] I.F. Akyildiz, W.Y. Lee, M.C. Vuran, and S. Mohanty," Next Generation/ Dynamic Spectrum Access/Cognitive Radio Wireless Networks: A Survey," *Computer Netw.*, vol. 50, pp. 2127–59, May 2006.

7 Optimal wavelet design for wireless communications

7.1 Introduction

Wavelet packet modulation (WPM) is an orthogonally multiplexed transmission technique where information-carrying symbols are placed on orthogonal wavelet packet bases [1]. These orthogonal signals are typically equi-spaced sub-carriers that are modulated to occupy different centre frequencies. Hence, this signalling is also referred to as multi-carrier modulation (MCM). In traditional implementations of MCM, such as the OFDM, the sub-carrier waveforms are Fourier bases or complex exponential functions. Recently, the wavelet packet transform has emerged as an important signal processing tool. The advantage is that WPM is a generic transmission scheme whose characteristics can be adjusted according to the system needs. The basis functions in wavelet packet representation are obtained from a single function called the mother wavelet through dilations (scaling) and translations (time-shifts). When the scales and translations are dyadic the resultant basis functions are orthogonal and span[1] embedded sub-spaces of $L^2(\mathbb{R})$,[2] at different resolutions yielding a multi-resolution analysis. From the perspective of communication system design, this has important and interesting implications – finite energy signals in $L^2(\mathbb{R})$ can be decomposed into orthogonal sub-spaces through a wavelet packet transform or conversely information can be packed into mutually orthogonal wavelet packet basis functions in a way that they do not interfere with one another. Since the basis functions and sub-spaces are orthogonal, they can be used in developing orthogonal waveforms for a wavelet-packet-based MCM leading to the idea of WPM.

In the previous chapters we evaluated the WPM system performance under metrics like PAPR (Chapter 5), and under loss of time/frequency/phase synchronization (Chapter 4). We also presented two applications of the WPM structure for estimation of spectrum and as a wideband multi-carrier modulation technology for dynamic spectrum access (Chapter 6). In this chapter we advance the state-of-the-art in WPM to design wavelet bases for use in communication formats. The possibility of adapting the characteristics of the WPM transmission is pursued with two *toy* examples where families

[1] The span of S may be defined as the collection of all (finite) linear combinations of the elements of S.

[2] Set of square-integrable functions in $\mathbb{R}$.

of wavelets, which are i) maximally frequency selective and ii) have the lowest cross-correlation energy, respectively, are developed. To this end a generic, unified framework that facilitates the design of new wavelet bases that cater to a requirement is established. Suitable optimizations are introduced when and where necessary to make the problem tractable. Numerical solvers are used to obtain the solution.

An important point to note is that by design of wavelet bases we essentially mean the design of filters used to obtain the wavelets. This is because the WPM system is realized with a tree structure made of cascaded half-band low/high-pass filter pairs. As we shall see later in the chapter, this is at once an advantage and a disadvantage. The advantage being that the design process is reduced to that of deriving finite impulse response (FIR) filters; hence standard, well-established methods can be employed; the disadvantage is that the relationship between the filters and the wavelet bases is sometimes not straightforward or explicit.

The rest of the chapter is organized as follows. Section 7.2 outlines the basics of the design process. The design process is exemplified with two examples in Sections 7.3 and 7.4. In Section 7.3 the design of maximally frequency selective wavelet is considered while Section 7.4 will delve into the filters with low cross-correlation errors. In each of these sections the design process is formulated as an optimization problem. The numerical results and their analysis is also presented in the respective sections. A summary of the material in the chapter is provided in Section 7.5.

7.2 Criteria for design of wavelets

7.2.1 Design procedure

The attributes of the WPM system greatly depend on the set of transmission bases utilized that in turn is determined by the filters used. This means that by adapting the filters one can adapt the WPM characteristics to satisfy a system specification. Choosing the right filter, though, is a delicate task. The filters cannot be arbitrarily chosen and instead have to satisfy a number of constraints. Besides the design objectives there are other *budgets* that have to be considered in order to guarantee that the designed wavelet is valid. The design procedure consists of three major steps, namely:

1. Formulation of design problem, i.e., stating the design objectives and constraints mandated by wavelet theory.
2. Application of suitable optimizations and transformations to make the problem tractable.
3. Utilization of numerical solvers to obtain the required filter coefficients.

At the end of the design procedure a low-pass FIR filter $h[n]$, satisfying the design and wavelet constraints, is obtained. From this filter the other three filters $g[n]$, $h'[n]$ and $g'[n]$, are derived through the QMF relation.

In the following sections we will elaborate on each of these processes.

7.2.2 Filter bank implementation of WPM

It is well known that compactly supported orthonormal wavelets can be obtained from a tree structure constructed by successively iterating discrete two-channel paraunitary filter banks [2]–[3]. Time- and frequency-limited orthonormal wavelet packet bases $\xi(t)$ can be derived by recursively iterating discrete half-band high-$g[n]$ and low-pass $h[n]$ filters, as[3]:

$$\begin{aligned} \xi_{l+1}^{2p}(t) &= \sqrt{2}\sum_{m} h[m]\xi_l^p(2t-m) \\ \xi_{l+1}^{2p+1}(t) &= \sqrt{2}\sum_{m} g[m]\xi_l^p(2t-m) \end{aligned} \tag{7.1}$$

In Eq. (7.1) the subscript l denotes the level in the tree structure and superscript p indicates the waveform index. The number of bases p generated is determined by the number of iterations l of the two-channel filter bank. Equation (7.1), known as a 2-scale equation, can be interpreted as follows – a basis function belonging to a certain sub-space of lower resolution can be obtained from shifted versions of the bases belonging to a sub-space of higher resolution; and the weights h and g used in the transformation are low and high pass in nature.

The filters h and g form a quadrature mirror pair and are also known as analysis filters. These filters have duals/adjoints known as synthesis filters that are also a pair of half-band low-h' and high-pass filters g'. All these four filters share a strict and tight relation and hence it is enough if the specifications of one of these filters are available. The wavelet packet sub-carriers (used at the transmitter end) are generated from the synthesis filters. And the wavelet packet duals (used at the receiver end) are obtained from the analysis filters. The entire WPM transceiver structure can thus be realized by this set of two QMF pairs. Hence, the design process can also be confined to the construction of one of the filters, usually the low-pass analysis filter h. A thorough analysis on the topic can be found in Chapters 2 and 3.

7.2.3 Important wavelet properties

The wavelet tool is a double-edged sword – on the one hand there is scope for customization and adaptation; on the other hand there are no clear guidelines to choose the best wavelet for a given application. In order to ease the selection process constraints, such as orthogonality, compact support and smoothness are imposed. We have outlined these properties in Chapter 2; here we shall discuss them in more detail.

7.2.3.1 Wavelet existence and compact support

This constraint is necessary to ensure that the wavelet has a finite non-zero coefficient and thus the impulse response of the wavelet decomposition filter is finite as well. According to [4], this property can be derived by simply integrating both sides of the

[3] The expressions are considered in the continuous-time domain to convenience derivations in Section 7.2.3.

two-scale equation in Eq. (7.1) and can be derived as follows[4]:

$$\int_{-\infty}^{\infty} \xi(t)dt = \sqrt{2}\int_{-\infty}^{\infty} \sum_{n} h[n]\xi(2t-n)dt$$
$$\int_{-\infty}^{\infty} \xi(t)dt = \sqrt{2}\sum_{n} h\,[n] \int_{-\infty}^{\infty} \xi(2t-n)dt \tag{7.2}$$
$$\int_{-\infty}^{\infty} \xi(t)dt = \sqrt{2}\sum_{n} h\,[n] \int_{-\infty}^{\infty} 0.5\xi(2t-n)d(2t-n)$$

Substituting $u = 2t - n$, Eq. (7.2) can be rewritten as:

$$\int_{-\infty}^{\infty} \xi(t)dt = \frac{1}{\sqrt{2}}\sum_{n} h\,[n] \int_{-\infty}^{\infty} \xi(u)du$$
$$\frac{\int_{-\infty}^{\infty} \xi(t)dt}{\int_{-\infty}^{\infty} \xi(u)du} = \frac{1}{\sqrt{2}}\sum_{n} h\,[n]$$

Finally, we obtain the compactly supported wavelet constraint as:

$$\sum_{n} h\,[n] = \sqrt{2} \tag{7.3}$$

It should be noted that the derivation that is given above is only possible if the scaling function is absolutely integrable and the integration of the scaling function is non-zero. Due to this fact, Eq. (7.3) is also recognized as the wavelet existence constraint.

7.2.3.2 Paraunitary condition

The paraunitary or the orthogonality condition is essential for many reasons. First, it is a prerequisite for generating orthonormal wavelets [2]–[3]. Secondly, it automatically ensures perfect reconstruction of the decomposed signal; i.e., the original signal can be reconstructed without amplitude or phase or aliasing distortion. To satisfy the paraunitary constraint the scaling filter coefficients have to be orthogonal at even shifts [2]–[3]. The constraint can be derived using the orthonormality property of the scaling function and its shifted version as follows:

$$\int_{-\infty}^{\infty} \xi(t)\xi(t-k)dt = \delta[k] \tag{7.4}$$

[4] The subscripts denoting the decomposition level l and the waveform index p have been dropped for convenience.

Substituting the two-scale Eq. (7.1) in Eq. (7.4) we get:

$$\int_{-\infty}^{\infty} \sum_{n} h\,[n]\,\xi(2t-n)\sqrt{2}\sum_{m} h\,[m]\,\xi(2(t-k)-m)\sqrt{2}dt = \delta[k]$$

$$2\sum_{n} h\,[n]\sum_{m} h\,[m]\int_{-\infty}^{\infty} \xi(2t-n)\xi(2(t-k)-m)dt = \delta[k] \tag{7.5a}$$

$$2\sum_{n} h\,[n]\sum_{m} h\,[m]\int_{-\infty}^{\infty} 0.5\xi(2t-n)\xi(2(t-k)-m)d(2t) = \delta[k]$$

or

$$\sum_{n} h\,[n]h\,[n-2k] = \delta[k] \text{ for } k = 0, 1, \ldots, (L/2)-1 \tag{7.5b}$$

Equation (7.5b) is called the double shift orthogonality relation of the wavelet low-pass filters impulse responses. In Eq. (7.5b), L illustrates the length of the low-pass wavelet filter impulse response. For a filter of length L the orthogonality condition (7.5b) imposes $^{L}/_{2}$ non-linear constraints on $h[n]$.

7.2.3.3 Flatness/*K*-regularity

This property is a rough measure of smoothness of the wavelet. The regularity condition is needed to ensure that the wavelet is smooth in both time and frequency domains [5]. It is normally quantified by the number of times a wavelet is continuously differentiable. The simplest regularity condition is the *flatness* constraint that is stated on the low-pass filter. A low-pass filter (LPF) is said to satisfy the Kth-order flatness if its transfer function $H(\omega)$ contains K zeroes located at the Nyquist frequency ($\omega = \pi$). For any function $Q(\omega)$ with no poles or zeros at ($\omega = \pi$), this can be written as:

$$H(\omega) = \left(\frac{1+e^{j\omega}}{2}\right)^{K} Q(\omega) \text{ with } Q(\pi) \neq 0 \tag{7.6}$$

In Eq. (7.6), $Q(\omega)$ is a factor of $H(\omega)$ that does not have any single zero at $\omega = \pi$. Having K number of zeros at $\omega = \pi$ also means that $H(\omega)$ is K-times differentiable and its derivatives are zero when they are evaluated at $\omega = \pi$. Considering that:

$$H(\omega) = \sum_{n} h\,[n]\exp(-j\omega n), \tag{7.7}$$

the kth-order derivative of $H(\omega)$ would be:

$$H^{(k)}(\omega) = \sum_{n} h\,[n]\,(-jn)^{k}\exp(-j\omega n) \tag{7.8}$$

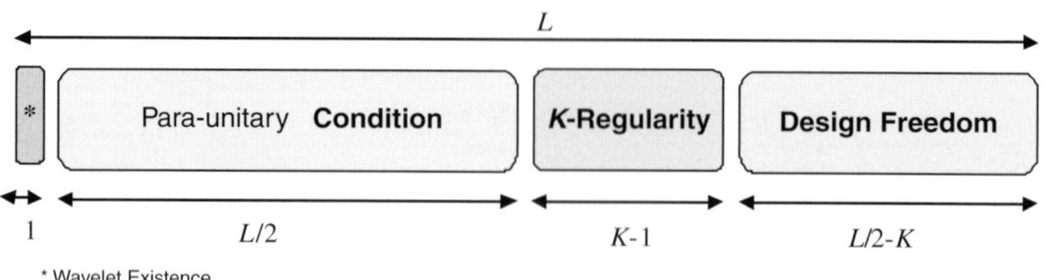

Figure 7.1 Wavelet conditions and degrees of freedom for design.

The evaluation of Eq. (7.8) at $\omega = \pi$ would result in:

$$H^{(k)}(\pi) = \sum_n h[n](-jn)^k \exp(-j\pi n)$$

$$\sum_n h[n](-j)^k (n)^k (e^{-j\pi})^n = 0$$

$$\sum_n h[n](-1)^n (n)^k = 0$$

Therefore, the K-regularity constraint in terms of the low-pass filter coefficients can be given as:

$$\sum_n h[n](n)^k(-1)^n = 0 \quad \text{for } k = 0, 1, 2, \ldots, K-1 \tag{7.9}$$

7.2.4 Degrees of freedom to design

The criteria (7.3), (7.5b) and (7.9) are necessary and sufficient conditions for the set to form an orthonormal basis and these conditions have to be imposed for all design procedures. For a filter of length L this is essentially getting L unknown filter variables from L equations. Of these L equations, one equation is required to satisfy the wavelet existence condition, $^L/_2$ come from the paraunitaryness constraint, $K-1$ from the regularity constraint and the remaining $^L/_2 - K$ conditions offer the possibility for establishing the design objective. The larger the value of $^L/_2 - K$, the greater the degree of freedom for design and the greater is the loss in regularity. There is therefore a trade-off on offer. The $^L/_2 - K$ degrees of freedom that remain after satisfying wavelet existence, orthogonality and K-regularity condition can be used to design a scaling filter with the desired property (refer to Figure 7.1). In Sections 7.3 and 7.4 we illustrate this with two examples.

7.3 Example 1 – Maximally frequency selective wavelets

As a first example we consider the design of filters that are maximally frequency selective. Frequency selectivity is a useful property for many applications, especially in the fields

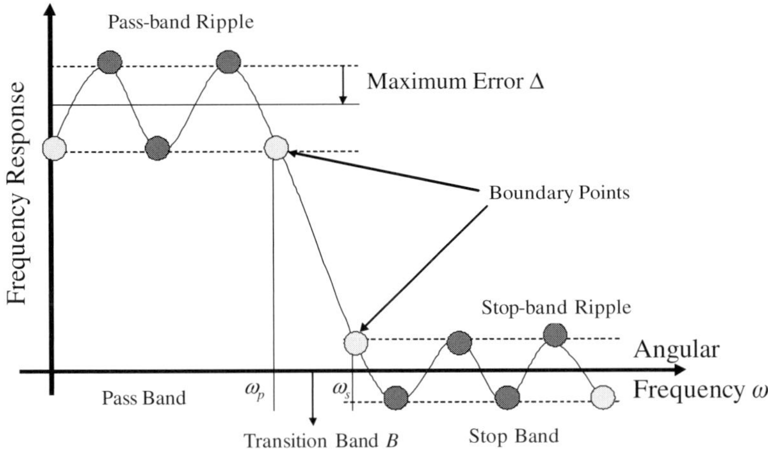

Figure 7.2 Plot of magnitude response $|H(\omega)|^2$ of the designed filter.

of cognitive radio, dynamic spectrum access and LTE-advanced[5], where the spectrum of a transmission signal has to be accurately shaped to match a frequency mask with low leakages to the neighbouring bands. As we shall see in Section 7.3.5, where the results are presented, the frequency selective filters yield wavelet bases with a well-confined spectral footprint. Such bases are ideal for the wavelet-packet-based spectrum estimator (WPSE) presented in Chapter 6.

To obtain the frequency selective filters, the design parameters are stated in the frequency domain in terms of the desired magnitude response $|H(\omega)|^2$ of the LPF (see Figure 7.2). In the figure ω_p and ω_s, denote pass- and stop-band frequencies, respectively $[0, \omega_p]$ is called the pass-band, $[\omega_s, \text{II}]$ is called the stop-band and $[\omega_p, \omega_s]$ is the transition band B_t. Δ_b connotes the maximum ripple that can be tolerated and the stop-band constraint can be stated as:

$$0 \leq |H(\omega)|^2 \leq \Delta_b \quad \text{for } \omega \in [\omega_s, \pi] \tag{7.10}$$

where

$$|H(\omega)|^2 = \sum_n \sum_m h[n]h[m]e^{-j\omega(n-m)} \tag{7.11}$$

In terms of the impulse response, (7.11) can be stated as:

$$0 \leq \sum_n \sum_m h[n]h[m]e^{-j\omega(n-m)} \leq \Delta_b \text{ for all } \omega \in [\omega_s, \pi] \tag{7.12}$$

[5] In long-term evolution advanced (LTE-advanced) the spectrum can be allocated over non-contiguous frequency bands. This possibility necessitates that the frequency bands are well confined without any side-lobes or spill over.

7.3.1 Formulation of design problem

This procedure was originally proposed by Parks and McClellan [6] for the design of FIR filters. However, it has to be adapted to accommodate the constraints (see Section 7.2.4) mandated by the wavelet theory [7]. The design goal is to generate filters with the desired transition band B_t and minimum error Δ_b while satisfying the wavelet constraints. For a given transition band B_t, this optimization problem can be formally stated as:

Problem 1: Minimize Δ_b subject to the wavelet constraints (7.3), (7.5b), and (7.9) *and the filter constraint* (7.12).

i.e.,

$$\text{MINIMIZE: } \Delta_b$$
$$\text{SUBJECT TO: } \begin{array}{l} \sum_n h[n] = \sqrt{2} \\ \sum_n h[n]h[n-2k] = \delta[k] \quad \text{for } k = 0, 1, \ldots, (L/2) - 1 \\ \sum_n h[n](n)^k(-1)^n = 0 \quad \text{for } k = 0, 1, 2, \ldots, K - 1 \\ 0 \le |H(\omega)|^2 \le \Delta_b \quad \text{for } \omega \in [\omega_s, \pi] \end{array} \tag{7.13}$$

for fixed values of transition band B_t, L and K.

It should be noted that we define the stop-band constraint only within the range of $\omega \in [\omega_s, \pi]$ due to the built in anti-symmetry of $(|H(\omega)|^2 - 1)$ about $\omega = \pi/2$ (see Figure 7.2), [7]. The stop-band constraint can be simplified further as follows:

$$\begin{aligned} |H(\omega)|^2 &= H(e^{j\omega})H(e^{-j\omega}) = \left(\sum_n h[n]e^{-j\omega n}\right)\left(\sum_m h[m]e^{j\omega m}\right) \\ |H(\omega)|^2 &= \sum_n \sum_m h[n]h[m]e^{-j\omega(n-m)} \end{aligned} \tag{7.14}$$

Hence, the stop-band constraint can be written as:

$$0 \le \sum_n \sum_m h[n]h[m]e^{-j\omega(n-m)} \le \Delta_b, \quad \forall \omega \in [\omega_s, \pi] \tag{7.15}$$

From Eqs. (7.5b) and (7.15), it is obvious that both double shift orthogonality and stop-band constraints are non-linear and non-convex. Therefore, the whole optimization problem is non-linear as well as non-convex. Therefore, the optimization problem as given above can only be solved by general-purpose solvers. However, such solvers are susceptible to being trapped in local minima. In order to overcome this difficulty, some authors have suggested multiple starting-point techniques or branch-and-bound method [8]. Moreover, general-purpose algorithms cannot guarantee that the found result is a global minimum and, furthermore, when the number of constraints increases these algorithms often fail to provide a valid solution.

The objective function and constraints can be solved much more efficiently using convex optimization and semi-definite programs [9]–[15][6]. In the following sections we

[6] In Appendix A1 we briefly discuss convex optimization and semi-definite programming.

attempt to express the design constraints in convex form so that convex optimization tools can be employed to obtain the solution [16]–[18].

7.3.2 Transformation of non-convex problem to linear/convex problem

Fortunately, it is possible to transform the non-convex/non-linear equations into a linear/convex problem by reformulating the constraints in terms of the autocorrelation sequence $r_h[k]$ [20]–[22]:

$$r_h[k] = \sum_{m \in z} h[m]h[m+k] \tag{7.16}$$

Taking into account the inherent symmetry of the autocorrelation sequence, it can be defined more precisely as:

$$r_h[l] = \sum_{n=0}^{L-l-1} h[n]h[n+l] \quad \text{for } l \geq 0 \tag{7.17}$$

In Eq. (7.17), L is the length of the FIR filter and the autocorrelation function is symmetric about $l = 0$; i.e.:

$$r_h[-l] = r_h[l] \tag{7.18}$$

The four constraints (7.3), (7.5b), (7.9) and (7.15) are derived in terms of $r_h[l]$ in the following sections.

7.3.2.1 Compact support or admissibility constraint

The compact support constraint in Eq. (7.3) can be rewritten as:

$$\sum_{n=0}^{L-1} h\,[n] = \sqrt{2} \text{ or,}$$

$$\sum_{n=0}^{L-1} h\,[n] \sum_{m=0}^{L-1} h\,[m] = 2.$$

Taking $m = n + l$, we have:

$$\sum_{n=0}^{L-1} \sum_{l=-n}^{L-n-1} h\,[n]\, h\,[n+l] = 2$$

Reversing the order of the summation and considering the fact that the impulse response of filter $h[n]$ has non-zero values only at $0 \leq n \leq L-1$, we obtain:

$$\sum_{l=-(L-1)}^{L-1} \sum_{n=0}^{L-l-1} h\,[n]\, h\,[n+l] = 2 \tag{7.19}$$

The compact support constraint in Eq. (7.3) can then be rewritten as:

$$\sum_{l=-(L-1)}^{L-1} r_h\,[l] = 2 \tag{7.20}$$

Taking into consideration the double shift orthonormality property (see Eq. (7.5b)) and the fact that the autocorrelation sequence is symmetric, we can simplify Eq. (7.20) further as:

$$\begin{aligned} r_h[0] + 2\sum_{l=1}^{L-1} r_h[l] &= 2 \\ \sum_{l=1}^{L-1} r_h[l] &= \frac{1}{2} \end{aligned} \tag{7.21}$$

Equation (7.21) is the compactly supported wavelet constraint stated in terms of the autocorrelation sequence $r_h[l]$.

7.3.2.2 Double shift orthogonality constraint

The double shift orthogonality constraint presented in Eq. (7.5b), can be expressed in terms of the autocorrelation sequence $r_h[l]$ as follows:

$$\sum_m h[m]h[m+2k] = r_h[2k] = \delta[k] \tag{7.22}$$

It should be noted that Eq. (7.22) is obtained by applying $n-2k=m$ on Eq. (7.5b). Hence, the final double shift orthogonality constraint in terms of autocorrelation sequence $r_h[l]$ is:

$$r_h[2k] = \delta[k] = \begin{cases} 1, & \text{for } k=0 \\ 0, & \text{otherwise} \end{cases} \quad \text{with } k = 0, 1, \ldots, \left\lfloor \frac{L-1}{2} \right\rfloor \tag{7.23}$$

Again, we make use of the symmetry property to simplify it. In contrast to Eq. (7.5b) which was non-convex, Eq. (7.23) consists of linear equalities and is also convex.

7.3.2.3 *K*-regularity constraint

The regularity constraint can be reformulated in terms of autocorrelation sequence $r_h[l]$ by considering the square of the absolute value of Eq. (7.6); i.e.:

$$|H(\omega)|^2 = \left(\frac{1+e^{-j\omega}}{2}\right)^K \left(\frac{1+e^{j\omega}}{2}\right)^K |Q(\omega)|^2 \tag{7.24}$$

Requiring the transfer function $H(\omega)$ to have K zeros at the Nyquist frequency ($\omega = \pi$) is equivalent to requiring $|H(\omega)|^2$ to have $2K$ zeros at $\omega = \pi$. Taking into account the fact that $|H(\omega)|^2$ is the Fourier transform of the autocorrelation sequence of $r_h[l]$, we can represent the $2k$th-order derivative of $|H(\omega)|^2$ as follows:

$$\left(|H(\omega)|^2\right)^{(2k)} = \sum_l r_h[l](-jl)^{2k} \exp(-j\omega l) \tag{7.25}$$

The evaluation of Eq. (7.25) at $\omega = \pi$ would result in:

$$\begin{aligned}\left(|H(\pi)|^2\right)^{(2k)} &= \sum_l r_h\,[l]\,(-jl)^{2k}\exp(-j\pi l)\\ 0 &= \sum_l r_h\,[l]\,(-j)^{2k}(l)^{2k}\left(\mathrm{e}^{-j\pi}\right)^l\\ \sum_l r_h\,[l]\,(l)^{2k}\,(-1)^l &= 0\end{aligned}\tag{7.26}$$

Now, for a filter of length L, the filter index l varies as $-(L-1) \le l \le (L-1)$, therefore Eq. (7.26) becomes:

$$\sum_{l=-L+1}^{L-1} (-1)^l\,(l)^{2k} r_h\,[l] = 0 \quad \text{for } k = 0, 1, \ldots, K-1 \tag{7.27}$$

In Eq. (7.27), K represents the desirable regularity index of the wavelet. Making use of the symmetry property of the autocorrelation sequence $r_h[l]$ and the fact that the term with $l = 0$ has zero value, Eq. (7.27) can be further simplified as:

$$\sum_{l=1}^{L-1} (-1)^l\,(l)^{2k} r_h\,[l] = 0 \quad \text{for } k = 0, 1, \ldots,\ K-1 \tag{7.28}$$

Equation (7.28) states the regularity constraint in terms of the autocorrelation sequence $r_h[l]$.

7.3.2.4 Stop-band constraint

Defining $n = m + k$, expression (7.15) can be written as:

$$|H(\omega)|^2 = \sum_m \sum_k h[m]h[m+k]e^{-j\omega(k)} = \sum_k r_h\,[k]\,e^{-j\omega k} \tag{7.29}$$

Therefore, the stop-band constraint becomes:

$$0 \le \sum_k r_h\,[k]\,e^{-j\omega k} \le \Delta_b \text{ for all } \omega \in [\omega_s, \pi] \tag{7.30}$$

The autocorrelation sequence $r_h[k]$ is symmetric about $k = 0$, (i.e., $r_h[l] = r_h[-l]$) [7]. Hence, Eq. (7.30) can be modified as:

$$\begin{aligned}|H(\omega)|^2 &= r_h\,[0] + \sum_l r_h\,[l]\left(e^{-j\omega l} + e^{j\omega l}\right)\\ &= r_h\,[0] + 2\sum_l r_h\,[l]\cos(\omega l) \quad \text{for } l = 1, 2, \ldots, L-1 \text{ and } \omega \in [\omega_s, \pi]\end{aligned}\tag{7.31}$$

Consequently, the stop-band constraint in Eq. (7.30) is written as:

$$0 \leq r_h[0] + 2\sum_l r_h[l]\cos(\omega l) \leq \Delta_b \quad \text{for } l = 1, 2, \ldots, L-1 \text{ and } \omega \in [\omega_s, \pi] \tag{7.32}$$

7.3.2.5 Spectral factorization and discretization on stop-band constraint

The reformulated optimization problem consists of the objective function and constraints expressed in terms of the autocorrelation sequence $r_h[l]$ and therefore the optimal solution will also be in the autocorrelation domain. Since our interest is the filter coefficients $h[n]$, we need to be able to obtain $h[n]$ from $r_h[l]$. There are no unique solutions to the filter coefficient that can be obtained for a given $r_h[l]$[7]. We borrow the spectral factorization algorithm proposed in [19] to obtain unique filters that satisfy the minimum-phase property[8] [20]. The spectral factorization of an autocorrelation sequence $r_h[l]$ can be performed as long as the logarithm function of its Fourier transform $R_h(\omega)$, which is simply $|H(\omega)|^2$, remains in $\mathbb{R}$[9]. To ensure this, the following additional constraint is enforced:

$$R_h(\omega) = |H(\omega)|^2 \geq 0, \quad \text{for } \omega \in [0, \pi]. \tag{7.33}$$

Using Eq. (7.31) the time-domain representation of Eq. (7.33) can be written as:

$$r_h[0] + 2\sum_l r_h[l]\cos(\omega l) \geq 0 \text{ for } l = 1, 2, \ldots, L-1 \text{ and } \omega \in [0, \pi] \tag{7.34}$$

Since we have an infinite number of inequalities in Eq. (7.34), we discretize it in the interval $\omega \in [0, \pi]$. This is necessary in order to make the optimization problem practically solvable. One such approach is proposed in [19] where the continuous variable ω is replaced with the discrete variable $\omega_i = i\pi/d$, defined on the finite set $i = [0, \ldots, d]$. A typical value of d suggested in [19] is $15n$. As a result, the constraint required for successful spectral factorization after applying the discretization process becomes:

$$r_h[0] + 2\sum_{l=1}^{L-1} r_h[l]\cos(i\pi l/d) \geq 0 \quad \text{for } i = 0, 1, \ldots, d \tag{7.35}$$

For clarity, hereon, we refer to Eq. (7.35) as the spectral factorization constraint.

As with the spectral factorization constraints, the number of stop-band constraints defined in Eq. (7.32) is also infinite. Hence, the stop-band constraints also have to be discretized to make the problem practically solvable. After the discretization, the stop-band constraints in Eq. (7.32) can be rewritten as:

$$0 \leq r_h[0] + 2\sum_{l=1}^{L-1} r_h[l]\cos(i\pi l/d) \leq \Delta_b \quad \text{for } i = \left\lceil \frac{\omega_s}{\pi} \right\rceil * d, \ldots, d \tag{7.36}$$

[7] In theory, infinite such filter solutions are possible.

[8] Minimum phase filters are guaranteed to be stable, and hence they have been chosen.

[9] See Appendix A2 for more details on the Kolmogorov spectral factorization algorithm.

The optimization problem in terms of the autocorrelation sequence $r_h[l]$ can thus be summarized as:

$$
\begin{aligned}
&\text{MINIMIZE: } \Delta_b \\
&\qquad \sum_{l=1}^{L-1} r_h\,[l] = \frac{1}{2} \\
&\qquad r_h[2k] = \delta[k] = \begin{cases} 1, & \text{for } k = 0 \\ 0, & \text{otherwise} \end{cases} \quad \text{where } k = 0, 1, \ldots, \left\lfloor \frac{L-1}{2} \right\rfloor \\
&\text{SUBJECT TO: } \sum_{l=1}^{L-1} (-1)^l\,(l)^{2k} r_h\,[l] = 0 \quad \text{for } k = 0, 1, \ldots, K-1 \\
&\qquad 0 \leq r_h\,[0] + 2\sum_{l=1}^{L-1} r_h\,[l] \cos(i\pi l/d) \leq \Delta_b \quad \text{for } i = \left\lceil \frac{\omega_s}{\pi} \right\rceil * d, \ldots, d \\
&\qquad r_h\,[0] + 2\sum_{l=1}^{L-1} r_h\,[l] \cos(i\pi l/d) \geq 0 \quad \text{for } i = 0, 1, \ldots, d.
\end{aligned}
\tag{7.37}
$$

This optimization problem is clearly linear and convex.

7.3.3 Reformulation of the optimization problem in the $Q(\omega)$ function domain

As can be noted from Section 7.3.2, the optimization problem is linear. Therefore, in principle any linear or convex programming tool can be used to solve this optimization problem. However, a numerical problem may arise for long filters (large values of L). This numerical problem is caused by the fact that the matrix of the linear system composed by regularity constraint in Eq. (7.28) becomes ill-conditioned when the values of L and K are large [7], [18]. In order to alleviate this the optimization problem is expressed in terms of the $Q(\omega)$ function, which is defined as:

$$
\begin{aligned}
|H(\omega)|^2 &= \left(\frac{1+e^{-j\omega}}{2}\right)^K \left(\frac{1+e^{j\omega}}{2}\right)^K |Q(\omega)|^2 \\
|H(\omega)|^2 &= \left(\frac{\left(1+e^{j\omega}\right)\left(1+e^{-j\omega}\right)}{4}\right)^K |Q(\omega)|^2 \\
|H(\omega)|^2 &= \left(\frac{(1+\cos(\omega))}{2}\right)^K |Q(\omega)|^2
\end{aligned}
\tag{7.38}
$$

The time-domain representation of Eq. (7.38) can be shown to be [18]:

$$
r_h[l] = 2^{-2K} \sum_{n=-K}^{K} \binom{2K}{n+K} r_q[l-n] \quad l = 0, 1, \ldots, L-1 \tag{7.39}
$$

Here, $r_q[l]$ is also an autocorrelation sequence. As with $r_h[l]$, the symmetry property also holds good for $r_q[l]$. The constraints now are redefined in terms of the autocorrelation sequence $r_q[l]$.

7.3.3.1 Compact support constraint

The property of compact support for wavelets is stated in terms of the autocorrelation sequence $r_q[l]$ by combining Eqs. (7.3) and (7.38) and taking $\omega = 0$. It can be noted that:

$$\begin{aligned} &\sum_n h\,[n] = H(\omega)|_{\omega=0} = \sum_n h[n]\exp(-j\omega n)\bigg|_{\omega=0} = \sqrt{2} \\ &|H(\omega)|^2\big|_{\omega=0} = 2 \end{aligned} \tag{7.40}$$

Substituting Eq. (7.38) into Eq. (7.40), we obtain:

$$\left\{ \left(\frac{1+e^{-j\omega}}{2}\right)^K \left(\frac{1+e^{j\omega}}{2}\right)^K |Q(\omega)|^2 \right\}\bigg|_{\omega=0} = 2$$

$$|Q(\omega)|^2\big|_{\omega=0} = 2$$

$$\left\{ r_q\,[0] + 2\sum_{l=1}^{L_q-1} r_q\,[l]\cos(\omega l) \right\}\bigg|_{\omega=0} = 2$$

We finally come up with the compactly supported wavelet constraint in terms of the autocorrelation sequence $r_q[l]$ as follows:

$$r_q\,[0] + 2\sum_{l=1}^{L_q-1} r_q\,[l] = 2 \tag{7.41}$$

7.3.3.2 Double shift orthogonality constraint

Based on Eqs. (7.23) and (7.39), the double shift orthogonality constraint in terms of the autocorrelation sequence $r_q[l]$ can be represented as:

$$\begin{aligned} &r_h[2l] = 2^{-2K} \sum_{n=-K}^{K} \binom{2K}{n+K} r_q[2l-n] = \delta[l],\, l = 0, 1, \ldots, \left\lfloor \frac{L-1}{2} \right\rfloor \\ &\sum_{n=-K}^{K} \binom{2K}{n+K} r_q[2l-n] = 2^{2K}\delta[l],\, l = 0, 1, \ldots, \left\lfloor \frac{L-1}{2} \right\rfloor \end{aligned} \tag{7.42}$$

Equation (7.42) defines the double shift orthogonality constraints in term of autocorrelation sequence $r_q[l]$.

7.3.3.3 Spectral factorization constraint

The easiest way to reformulate the spectral factorization constraint in terms of the autocorrelation sequence $r_q[l]$ is by combining Eqs. (7.33) and (7.38) as follows:

$$\begin{aligned} &|H(\omega)|^2 \geq 0 && \text{for } \omega \in [0, \pi] \\ &\left(\frac{1+\cos(\omega)}{2}\right)^K |Q(\omega)|^2 \geq 0 && \text{for } \omega \in [0, \pi] \end{aligned} \tag{7.43}$$

Since the term $1 + \cos(\omega)$ in Eq. (7.43) is always positive, it can be rephrased as:

$$|Q(\omega)|^2 \geq 0 \quad \text{for } \omega \in [0, \pi] \tag{7.44}$$

Discretizing it in the interval $\omega \in [0, \pi]$, the spectral factorization constraint in terms of autocorrelation sequence $r_q[l]$ can be written as:

$$r_q[0] + 2\sum_{l=1}^{L_q-1} r_q[l]\cos(i\pi l/d) \geq 0 \quad \text{for } i = 0, 1, \ldots, d \tag{7.45}$$

It is clear from Eq. (7.6) that since $Q(\omega)$ has K zeros less than $H(\omega)$, the length of the filter $q[n]$ would be $L_q = L - K$.

7.3.3.4 Stop-band constraint

As with the spectral factorization constraints, the stop-band constraints in terms of autocorrelation sequence $r_q[l]$ is obtained by combining Eqs. (7.14), (7.15) and (7.38) as follows:

$$\begin{aligned} 0 &\leq |H(\omega)|^2 \leq \Delta_b && \text{for } \omega \in [\omega_s, \pi] \\ 0 &\leq \left(\frac{1+\cos(\omega)}{2}\right)^K |Q(\omega)|^2 \leq \Delta_b && \text{for } \omega \in [\omega_s, \pi] \end{aligned}$$

Discretizing it in the interval $\omega \in [\omega_s, \pi]$, the stop-band constraint can be expressed in terms of the autocorrelation sequence $r_q[l]$ as:

$$0 \leq \left(\frac{1+\cos(i\pi/d)}{2}\right)^K \left(r_q[0] + 2\sum_{l=1}^{L_q-1} r_q[l]\cos(i\pi l/d)\right) \leq \Delta_b$$
$$\text{for } i = \left\lceil \frac{\omega_s}{\pi} \right\rceil * d, \ldots, d \text{ and } L_q = L - K. \tag{7.46}$$

It is clear from Eq. (7.46) that when the optimization problem is expressed in terms of the autocorrelation sequence $r_q[l]$, the necessity for $|H(\omega)|^2$ to have $2K$ zeros at $\omega = \pi$ has been imposed implicitly. Therefore, the regularity constraints are not explicitly expressed when the optimization problem is conducted in the $Q(\omega)$ domain.

The spectral factorization constraints stated in Eqs. (7.43) and (7.45) will be automatically satisfied if the stop-band constraint stated in Eq. (7.46) is satisfied. In fact, the stop-band constraint (7.46) is more stringent than the spectral factorization constraint (7.45).

In summary, the optimization problem in terms of the autocorrelation sequence $r_q[l]$ can be stated as:

MINIMIZE: Δ

SUBJECT TO:

$$\begin{aligned} & r_q[0] + 2\sum_{l=1}^{L_q-1} r_q[l] = 2 \\ & \sum_{n=-K}^{K} \binom{2K}{n+K} r_q[2l-n] = 2^{2K}\delta[l] \quad \text{for } l = 0, 1, \ldots, \left\lfloor \frac{L-1}{2} \right\rfloor \\ & 0 \leq \left(\frac{1+\cos(i\pi/d)}{2}\right)^K \left(r_q[0] + 2\sum_{l=1}^{L_q-1} r_q[l]\cos(i\pi l/d)\right) \\ & \leq \Delta \quad \text{for } i = \left\lceil \frac{\omega_s}{\pi} \right\rceil * d, \ldots, d \text{ and } L_q = L - K. \end{aligned} \tag{7.47}$$

Once we find the optimal autocorrelation sequence $r_q[l]$, the spectral factorization is employed in order to derive the optimal sequence $q[l]$ from $r_q[l]$. Finally, the optimal wavelet low-pass filter coefficients are computed using the time domain equivalent of Eq. (7.6), [18]:

$$h[l] = 2^{-K} \sum_{k=0}^{K} \binom{K}{k} q\,[l-k] \tag{7.48}$$

7.3.4 Solving the convex optimization problem

Since the optimization problem posed above is linear it is also convex. Therefore, any linear or convex optimization tool can be used to solve this problem. In this case, we choose SeDuMi [23] as generic semi-definite programming (SDP) solvers to solve the optimization problem. SeDuMi stands for *self-dual minimization* as it implements a self-dual embedding technique for optimization over self-dual homogeneous cones [23]. It comes as an additional Matlab® package and can be used for linear, quadratic and semi-definite programming. Normally, it requires a problem to be described in a primal standard form but with modelling languages like YALMIP (short for yet another LMI parser) the optimization problems can be directly expressed in a user-friendly higher-level language [24]. Thus, YALMIP allows the user to concentrate on the high-level modelling without having to worry about low-level details. We have developed a filter optimization program that incorporates most of the available optimization routines for Matlab® and that relies on YALMIP to translate the problem into the standard form. The blocks of the filter design program are elucidated in Figure 7.3. The design process consists of both analytical and numerical modules. In the analytical part, the non-convex problem is converted into a convex one, followed by a transformation of the expression from the autocorrelation $r_h[l]$ domain into the autocorrelation $r_q[l]$ domain. In the numerical part the convex problem is solved and the solution obtained in terms of $r_q[n]$. After that, another analytical process is initiated to derive optimum low-pass filter coefficients $h[n]$ from the sequences $q[n]$, which is obtained by applying spectral factorization on $r_q[l]$. We use the spectral factorization algorithm proposed in [19]. From the autocorrelation sequence, this spectral factorization algorithm derives filter coefficients with length L having minimum phase property[10]. At the end of the design process the filter coefficients of the analysis LPF will be generated. From the analysis LPF $h[n]$, the HPF $g[n]$ and the synthesis filters, LPF $h'[n]$ and HPF $g'[n]$, can be obtained through the QMF equations. And from these sets of filters the WPM carriers and their duals can be derived using the 2-scale Eq. (7.1).

7.3.5 Results and analysis

In this section we present a few results to demonstrate the design procedure [25]–[28]. The main variables of the design process are the length and regularity order of the

[10] We chose filters having minimum phase because they guarantee stability.

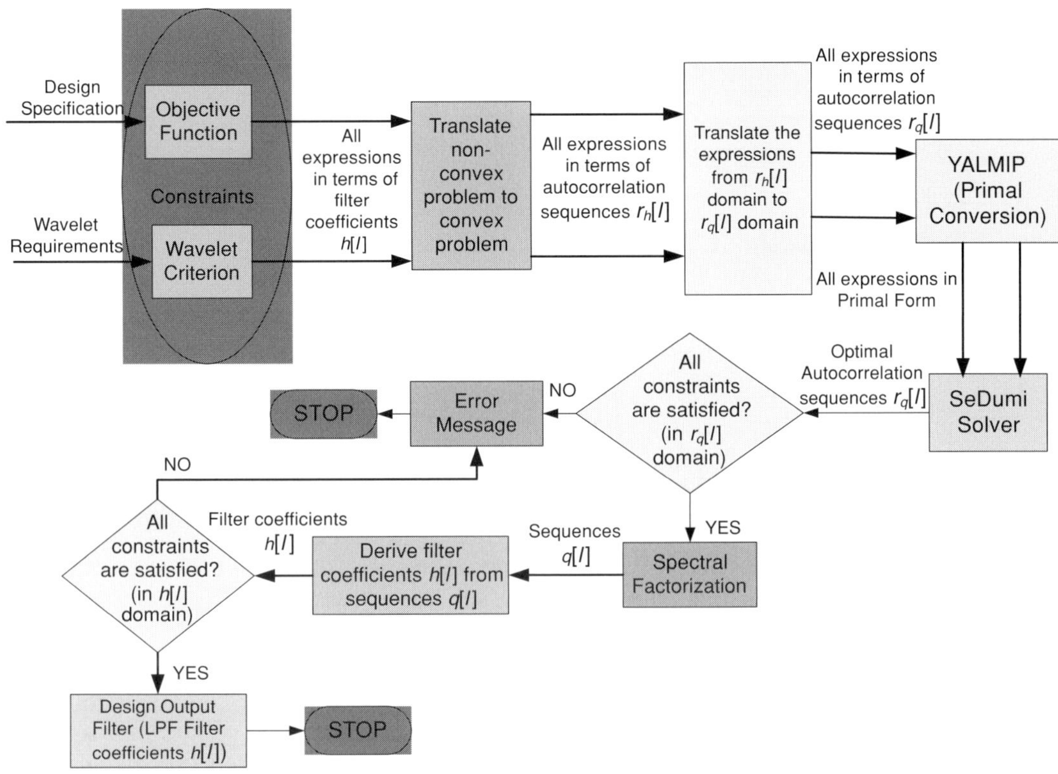

Figure 7.3 Flow chart of the optimum wavelet design process for wavelet-packet-based spectrum estimation.

filter. Regularity has to be equal to or larger than 1 to ensure that the wavelet existence constraint is satisfied and it may not exceed $^L/_2$. If the selected value for regularity is close to the upper limit, fewer degrees of freedom will be left for the optimization of the objective function. On the other hand, imposing a small regularity can result in highly irregular wavelets.

7.3.5.1 Frequency and impulse response of designed filter

We consider two wavelets with filter lengths $L = 30$ and $L = 40$. Indeed it is possible to design filters of other lengths, too. In the first example, shown in Figure 7.4, the frequency response of the designed wavelet filters is compared with Daubechies and Coiflet wavelet filters. Figure 7.5 compares the frequency response of the proposed wavelet filters to Daubechies and Coiflet wavelet filters. For fairness of comparison all of these wavelet filters have a filter length of 30 and 40. A K-regularity index of 7 and transition band (B_t) of 0.2π is enforced on the designed wavelet filters. From Figure 7.4, it is evident that the filters obtained from the design have better frequency selectivity, with sharper transition between the pass- and stop-bands, Daubechies and Coiflet counterparts. A small price, however, is paid in terms of the ripples introduced in the side-lobe. Figure 7.5 presents

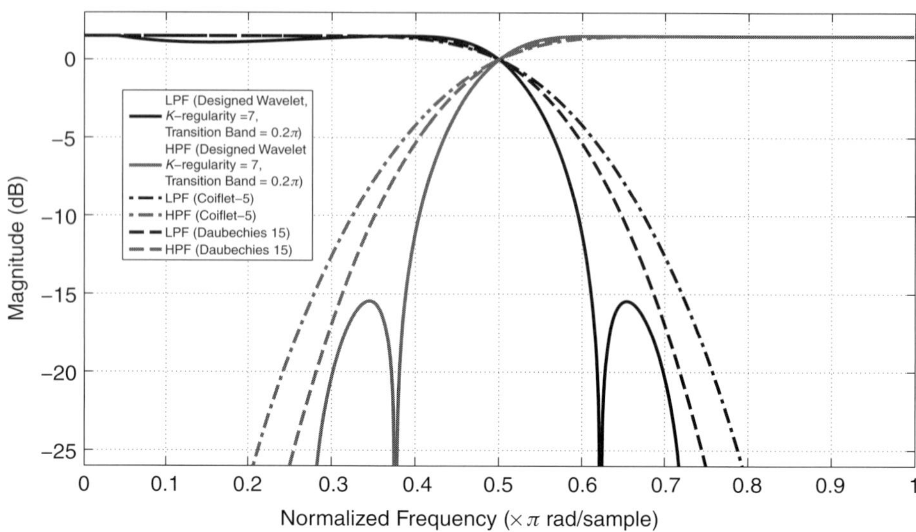

Figure 7.4 Frequency response of Daubechies 15, Coiflet 5 and the designed wavelet low-pass (LPF) and high-pass (HPF) filter with $L = 30$, $K = 7$, $B_t = 0.2\pi$.

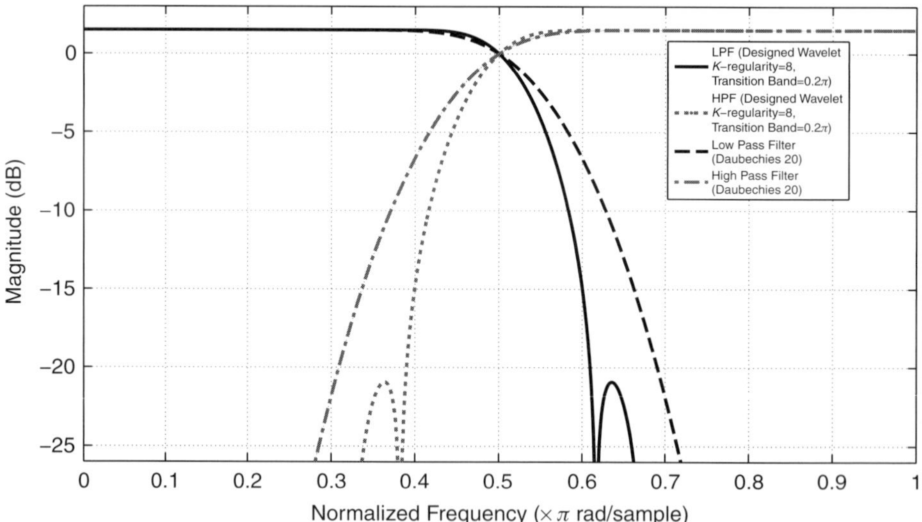

Figure 7.5 Frequency response of Daubechies 20 and the designed wavelet low-pass (LPF) and high-pass (HPF) filter with $L = 40$, $K = 8$, $B = 0.2\pi$.

similar comparison for $L = 40$. In this example, only the maximally frequency selective wavelet and Daubechies 20 filters are considered.

Figures 7.6 and 7.7 depict the impulse responses of the high- and low-pass filters of the optimally designed wavelets for $L = 30$, $K = 7$, $B_t = 0.2\pi$ and $L = 40$, $K = 8$, $B_t = 0.2\pi$, respectively. The coefficients of the designed wavelet filter for $L = 30$,

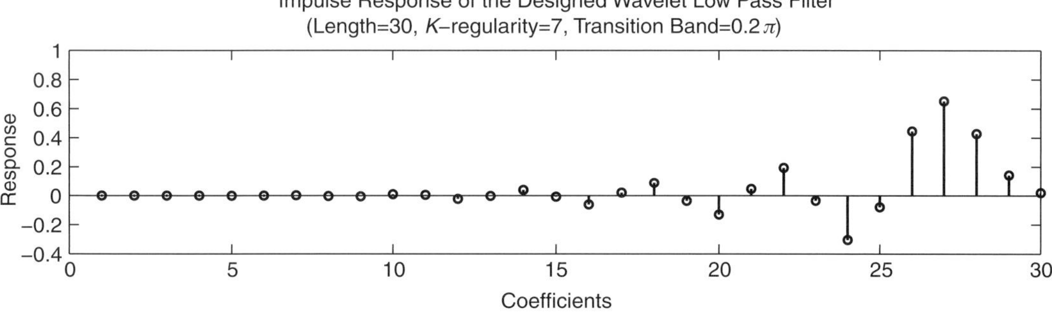

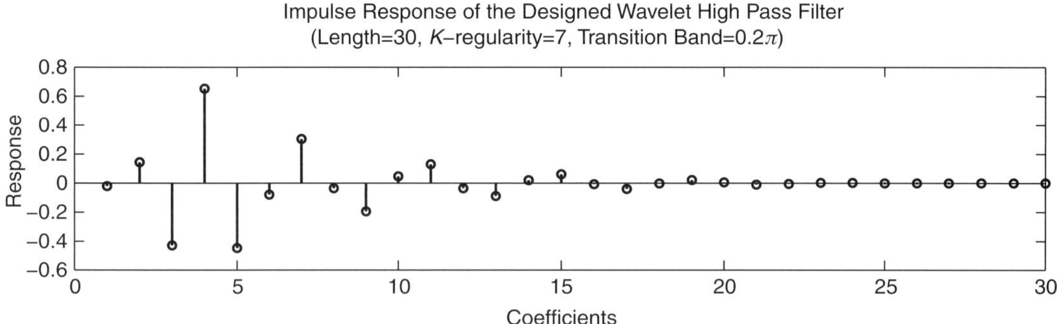

Figure 7.6 Impulse response of the designed optimal wavelet filter with length $L = 30$, K-regularity $K = 7$, overall transition band $B_t = 0.2\pi$.

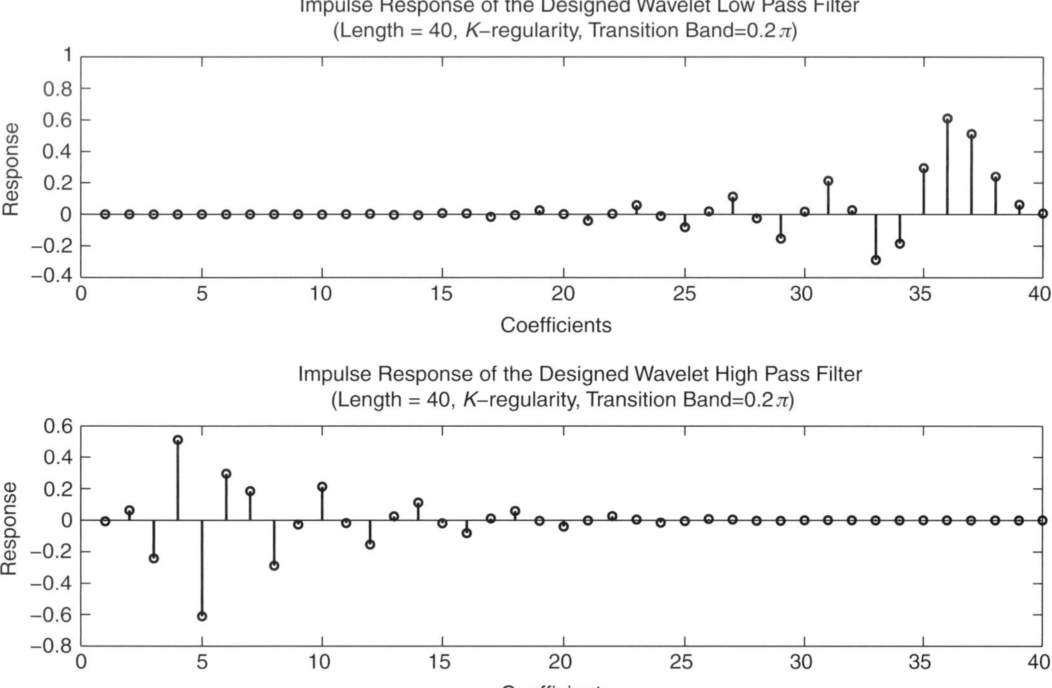

Figure 7.7 Impulse response of the designed optimal wavelet filter with length $L = 40$, K-regularity $= 8$, overall transition band $B_t = 0.2\pi$.

Table 7.1 Optimal filter coefficients for filter length $L = 30$, K-regularity $= 7$, transition band $= 0.2\pi$

Index	Low-Pass Filter	High-Pass Filter	Index	Low-Pass Filter	High-Pass Filter
1	0.0000	–0.0201	16	–0.0611	–0.0074
2	–0.0000	0.1437	17	0.0206	–0.0395
3	0.0001	–0.4279	18	0.0892	–0.0017
4	0.0002	0.6521	19	–0.0357	0.0223
5	–0.0006	–0.4454	20	–0.1296	0.0055
6	0.0001	–0.0789	21	0.0469	–0.0097
7	0.0024	0.3037	22	0.1943	–0.0049
8	–0.0026	–0.0350	23	–0.0350	0.0026
9	–0.0049	–0.1943	24	–0.3037	0.0024
10	0.0097	0.0469	25	–0.0789	–0.0001
11	0.0055	0.1296	26	0.4454	–0.0006
12	–0.0223	–0.0357	27	0.6521	–0.0002
13	–0.0017	–0.0892	28	0.4279	0.0001
14	0.0395	0.0206	29	0.1437	0.0000
15	–0.0074	0.0611	30	0.0201	0.0000

$K = 7$, $B_t = 0.2\pi$ and $L = 40$, $K = 8$, $B_t = 0.2\pi$ are presented in Tables 7.1 and 7.2, respectively.

7.3.5.2 Evaluation of spectrum estimator performance

Now, we examine the performance of the wavelet-packet based spectrum estimator or WPSE (detailed in Chapter 6) with the newly designed wavelet. For this purpose, three types of sources are considered, namely, partial band, single tone and multi-band. The partial-band source has its energy spread over a continuous range of frequencies and it occupies the normalized frequency band from 0.25π to 0.75π. The single-tone source has all of its energy at one frequency and is in the middle of the range spanned by wavelet-based spectrum estimation at 0.5π. The third source has a multi-band characteristic with three active bands occupying the normalized frequency bands of 0.08π–0.19π, 0.47π–0.58π, and 0.86π–0.97π, respectively. The details of all the sources are provided in Table 9.3.

Partial band source

Figure 7.8 presents how spectrum estimation of a partial band source with the newly designed wavelet compares with those based on standard wavelet family. Here, the number of samples is set to 12 800. The specifications for the optimal wavelet are L (length) $= 30$, K (regularity index) $= 7$ and B_t (transition bandwidth) $= 0.2\pi$. It is clear from the figure that the newly designed wavelet outperforms Daubechies, Coiflet and Symlet wavelets of the same length. The improvements are with regard to frequency

Table 7.2 Optimal filter coefficients for filter length $L = 40$, K-regularity $= 8$, transition band $= 0.2\pi$

Index	Low-Pass Filter	High-Pass Filter	Index	Low-Pass Filter	High-Pass Filter
1	0.0000	–0.0071	21	–0.0404	–0.0018
2	0.0000	0.0630	22	0.0035	0.0261
3	0.0000	–0.2416	23	0.0586	0.0047
4	0.0000	0.5128	24	–0.0110	–0.0154
5	0.0001	–0.6110	25	–0.0818	–0.0053
6	–0.0000	0.2958	26	0.0194	0.0077
7	–0.0001	0.1849	27	0.1122	0.0044
8	0.0004	–0.2882	28	–0.0249	–0.0030
9	0.0001	–0.0275	29	–0.1538	–0.0028
10	–0.0013	0.2128	30	0.0179	0.0006
11	0.0006	–0.0179	31	0.2128	0.0013
12	0.0028	–0.1538	32	0.0275	0.0001
13	–0.0030	0.0249	33	–0.2882	–0.0004
14	–0.0044	0.1122	34	–0.1849	–0.0001
15	0.0077	–0.0194	35	0.2958	0.0000
16	0.0053	–0.0818	36	0.6110	0.0001
17	–0.0154	0.0110	37	0.5128	–0.0000
18	–0.0047	0.0586	38	0.2416	0.0000
19	0.0261	–0.0035	39	0.0630	–0.0000
20	0.0018	–0.0404	40	0.0071	0.0000

Table 7.3 Description of four types of sources used in the experiments

	Type of sources	Description
1	Partial band	• Frequency occupied: $[0.25\pi, 0.75\pi]$
2	Single tone	• Frequency occupied: 0.5π
3	Multi-band	• Consist of 3 active bands occupying normalized frequency bands $[0.08\pi, 0.19\pi]$, $[0.47\pi, 0.58\pi]$, and $[0.86\pi, 0.97\pi]$, respectively.

selectivity and the sharp transition between the occupied band and the unoccupied band.

Single-tone source

On the other hand for the estimation of single-tone source, as illustrated by the plots in Figure 7.9, the difference in performances of the designed wavelet and the standard ones is not tangible. The frequency resolution of the single-tone source is influenced more by the levels of decomposition than to the frequency selectivity of the filter used. Hence, there is no perceivable differences in the performances of various wavelets.

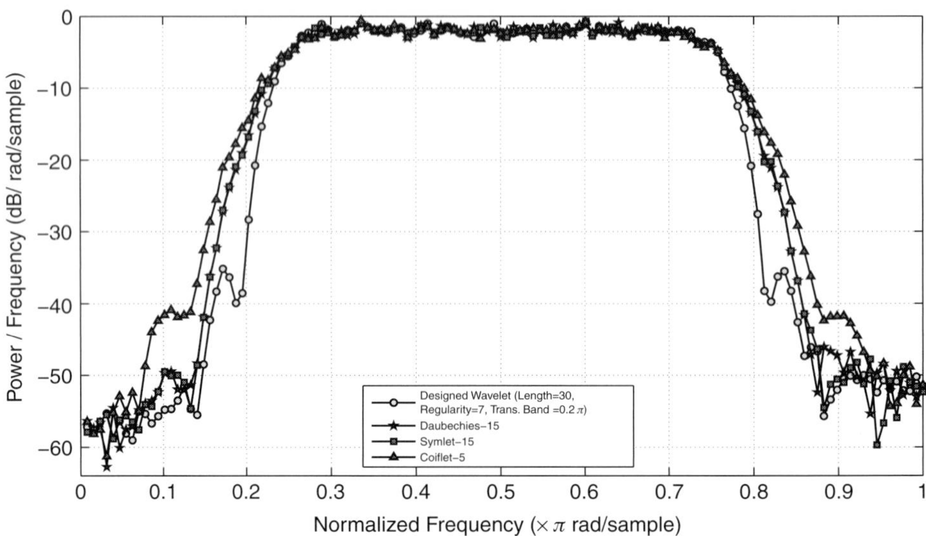

Figure 7.8 Estimates of partial band source based on Coiflet 5, Daubechies 15, Symlet 15 and the designed optimal wavelet filter with length $L = 30$, K-regularity $= 7$, overall transition band $B_t = 0.2\pi$. The wavelet decomposition level used here is 7. The number of samples in this experiment is 12 800.

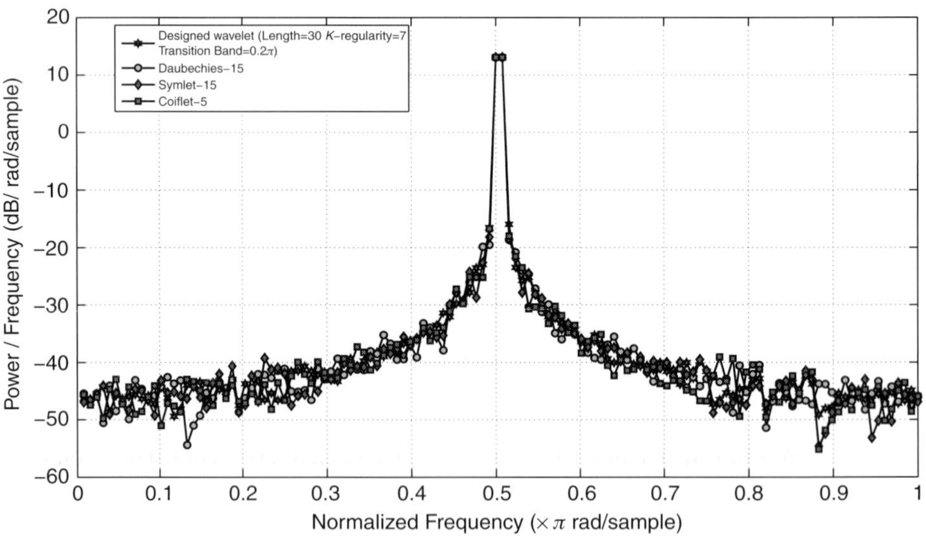

Figure 7.9 Estimates of single-tone source based on Coiflet 5, Daubechies 15, Symlet 15 and the designed optimal wavelet filter with length $L = 30$, K-regularity $= 7$, overall transition band $B_t = 0.2\pi$. The wavelet decomposition level used here is 7. The number of samples in this experiment is 12 800.

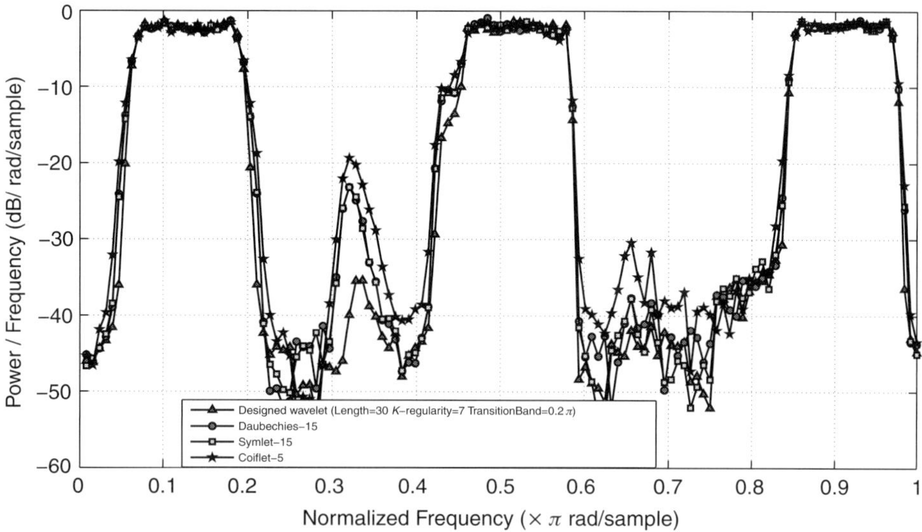

Figure 7.10 Spectrum shaping with WPM carriers based on the wavelets: Coiflet 5, Daubechies 15, Symlet 15 and the designed optimal wavelet filter with length $L = 30$, K-regularity $= 7$, transition band $B_t = 0.2\pi$.

Multiple-band source

The benefit of frequency selective filters is that the WPM carriers derived from them have narrow and well-confined spectral footprints. Moreover, they also aid in better estimation of signals. Figure 7.10 illustrates this characteristic where the frequency selective wavelets are shown to efficiently carve the bands between the desirable and undesired footprints, while all other wavelets have residual infringing components. This feature is useful in applications such as cognitive radio and LTE-advanced where the transmission signal characteristics have to be shaped to accurately map a frequency mask.

7.3.5.3 Evaluation of receiver operating characteristic

The receiver operating characteristic (ROC) is used as the second figure-of-merit to better gauge the system performance of the spectrum estimator. To obtain the probability of detection (P_d) and false alarm (P_{fa}), we divide the normalized frequency range $[0,\pi]$ into 128 equal bands (or frequency bins). Each bin is occupied by 1 source, meaning that overall there are 128 sources. These 128 sources are randomly activated/deactivated and the P_d and P_{fa} are calculated for each threshold for a sample space of 100 experiments. An active source operates around –2.1 dB power and the threshold is varied between –3 dB to –15 dB. The number of samples used to estimate is 12 800. Figure 7.11 depicts the P_d and P_{fa} as a function of threshold level; the plots clearly underline the superiority of the newly designed wavelet in relation to other wavelet families of the same filter length. The frequency selectivity inherent in the proposed wavelet has allowed spectrum estimator built on it to have much better P_d and P_{fa} for all thresholds in comparison to

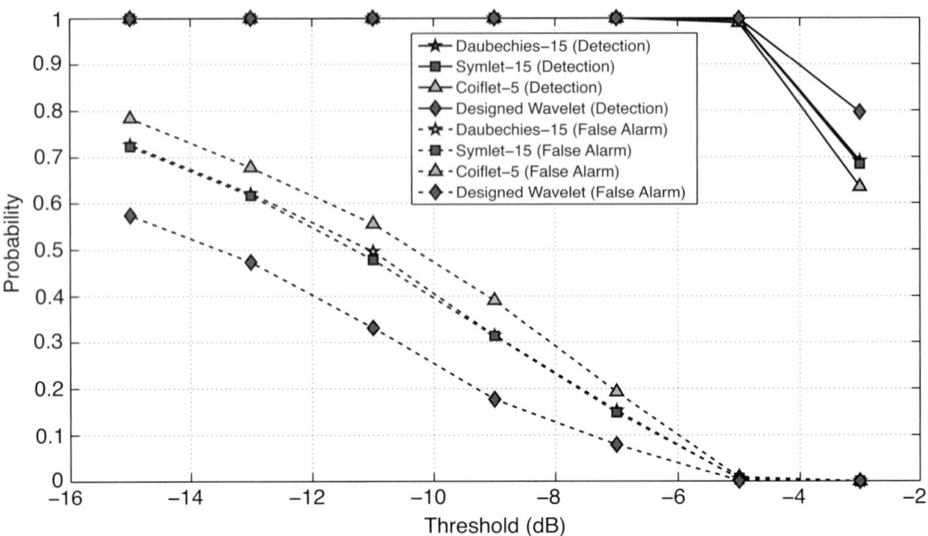

Figure 7.11 Detection and false alarm probability of spectrum estimation based on various wavelet families. In this scenario, the length of the wavelet decomposition filter is 30, the wavelet decomposition level is 7 and the sample space is of size 12 800. The K-regularity of the designed wavelets with SDP is 7 with a normalized transition band B_t of 0.2π.

Daubechies-, Symlet- and Coiflet-based estimators. The ROC depicted in Figure 7.12 further enhances the benefits the superiority of the estimator based on the designed wavelet.

7.3.5.4 Other studies – Filter characteristics and their influence

We now study the impact of altering the filter design parameters on the ROC. The plots in Figure 7.13 show the impact of filter length on the ROC. The results shows that for a given regularity order, the longer the filters, the better the P_d and P_{fa} of the estimates. This is on expected lines because filters that are longer offer more degrees of freedom to minimize the pass-band and stop-band ripple. Likewise, for a given filter length, a lower K regularity index results in more degrees of freedom available to minimize the pass-/stop-band ripple yielding better performance results.

Figure 7.14 exemplifies the influence of transition-band variation on the detection and false alarm probability. The result further exemplifies the importance of frequency selectivity on the quality of the estimates. Here, configurations with narrower transition bands offer lower false alarm and higher detection probability.

7.4 Example 2 – Wavelets with low cross-correlation error

As a second example we design filters with low cross-correlation energy between the low- and high-pass filters with the objective of minimizing the interference due to timing

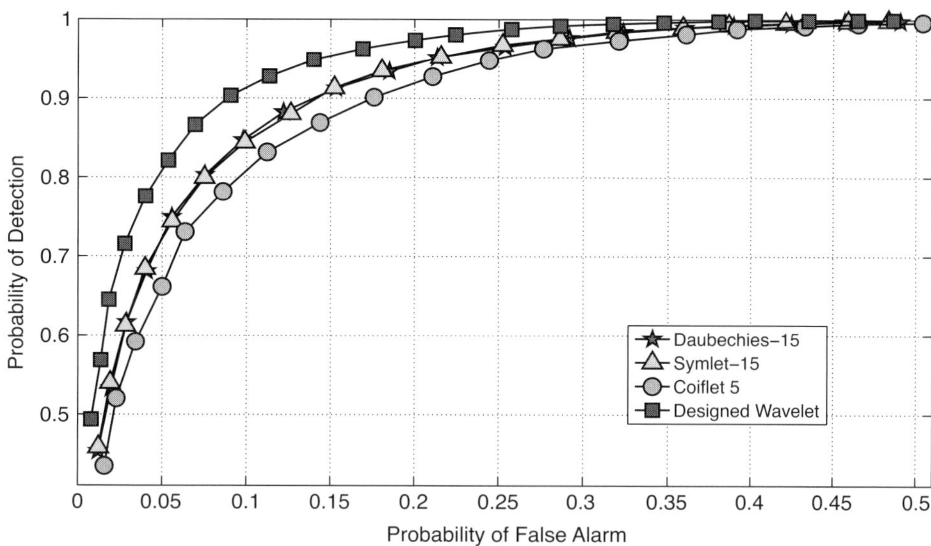

Figure 7.12 Receiver operating characteristic of spectrum estimation based on various wavelet families. In this scenario, the length of the wavelet decomposition filter is 30, the wavelet decomposition level is 7 and the sample space is of size 384. The K-regularity of the designed wavelets with SDP is 7 with a normalized transition band of 0.2π.

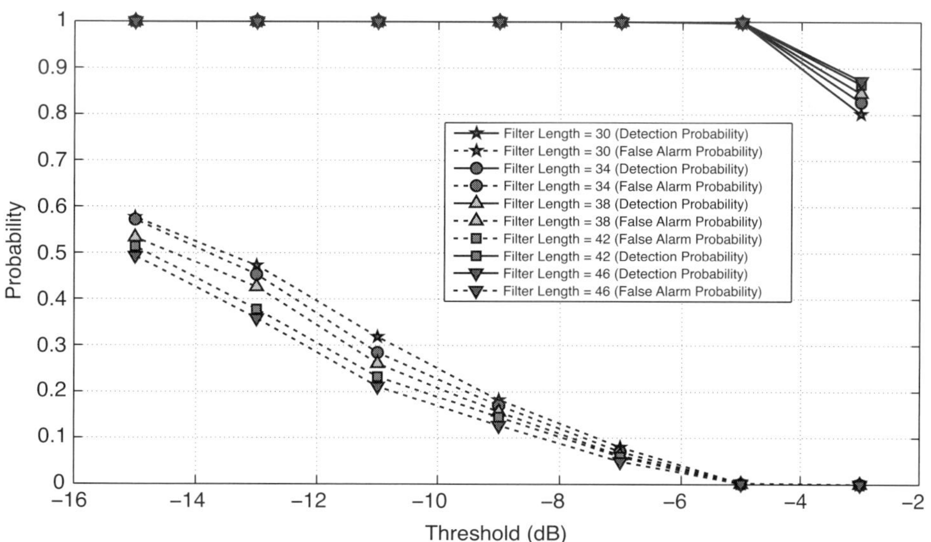

Figure 7.13 Detection and false alarm probability of spectrum estimation based on newly designed wavelet with variations on filter lengths. In this scenario, the wavelet decomposition level is 7 and the sample space is of size 12 800. The K-regularity of the designed wavelets with semi-definite programming (SDP) is 7 with a transition band of 0.2π.

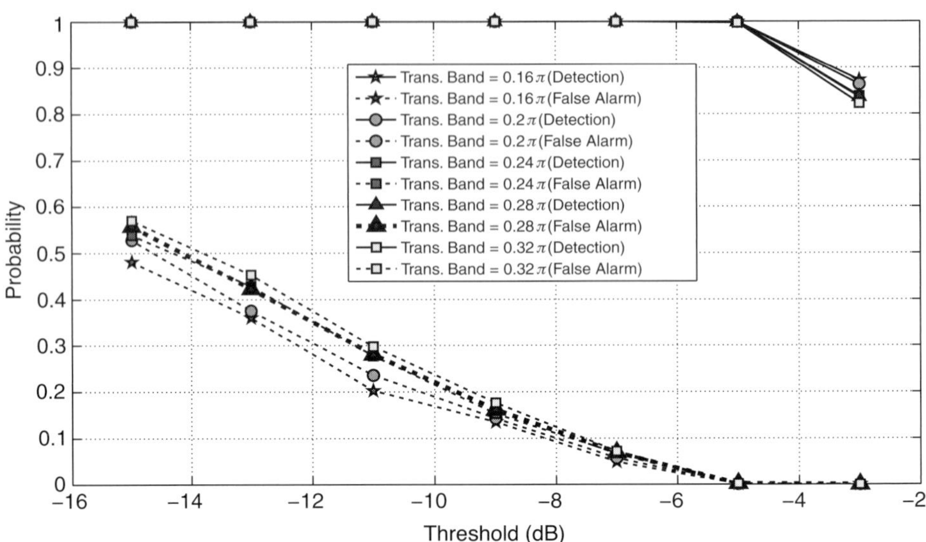

Figure 7.14 Detection and false alarm probability of spectrum estimation based on newly designed wavelet with variations on transition band. In this scenario, the length of the wavelet decomposition filter is 40, the wavelet decomposition level is 7 and the sample space is of size 12 800. The K-regularity of the designed wavelets with semi-definite programming (SDP) is 6.

error in WPM transmission. In Chapter 4 we found out that multi-carrier systems are highly sensitive to loss of time synchronization. A loss of time synchrony results in samples outside a WPM symbol getting selected erroneously, while useful samples at the beginning or at the end of the symbol are discarded. It also introduces ISI and ICI, causing performance degradation.

We also observed that though WPM and OFDM share many similarities as orthogonal multi-carrier systems, they are significantly different in their responses to loss of time synchronization. This difference is a result of the fact that the WPM symbols overlap with each other and are longer than the OFDM symbol[11]. Under a loss in time synchronization, the overlap of the symbols in WPM causes each symbol to interfere with several other symbols, while in OFDM each symbol interferes only with its neighbours. The second important difference is in the usage of guard intervals. OFDM benefits from the cyclic prefix that significantly improves its performance under timing errors. WPM cannot use guard intervals because of the symbol overlap.

Fortunately, WPM offers the possibility of adjusting the properties of the waveforms in a way that the errors due to loss of synchronization can be minimized. In this section we present a method to design a new family of wavelet filters that minimize the energy of the timing error interference.

[11] The length of the symbol and the degree of overlap are determined by the length of wavelet filter used.

7.4.1 Time offset errors in WPM

The time synchronization error is modelled by shifting the received data samples $R[n]$ by a time offset Δ_t to the left or right as:

$$R[n \pm \Delta_t] = S[n] + w[n] \tag{7.49}$$

Here, $S[n]$ denotes the transmitted signal and $w[n]$ the Gaussian noise.

Recalling, from Chapter 3, that under ideal conditions, when the WPM transmitter and receiver are perfectly synchronized and the channel is benign, the estimation of the data contained in the uth symbol and kth sub-carrier $\hat{a}_{u',k'}$ are the same as the transmitted data $a_{u,k}$[12]. However, errors are introduced in the demodulation decision-making process under time offset errors Δ_t as elucidated below:

$$\begin{aligned}\hat{a}_{u',k'} &= \sum_n R[n]\xi_l^{k'}[u'N - n + \Delta_t] = \sum_n \sum_u \sum_{k=0}^{N-1} a_{u,k}\xi_l^k[n - uN]\xi_l^{k'}[u'N - n + \Delta_t] \\ &= \sum_u \sum_{k=0}^{N-1} a_{u,k}\left(\sum_n \xi_l^k[n - uN]\xi_l^{k'}[u'N - n + \Delta_t]\right)\end{aligned} \tag{7.50}$$

Defining the cross-waveform function $\Omega(\Delta_t)$ as:

$$\Omega_{k,k'}^{u,u'}[\Delta_t] = \sum_n \xi_l^k[n - uN]\xi_l^{k'}[u'N - n + \Delta_t] \tag{7.51}$$

the demodulated data corrupted by the interference due to loss of orthogonality at the receiver for the kth sub-carrier and uth symbol can be expressed as:

$$\hat{a}_{u',k'} = \underbrace{a_{u',k'}\Omega_{k',k'}^{u',u'}[\Delta_t]}_{\text{Desired Alphabet}} + \underbrace{\sum_{u;u\neq u'} a_{u,k'}\Omega_{k',k'}^{u,u'}[\Delta_t]}_{\text{ISI}} + \underbrace{\sum_u \sum_{k=0;k\neq k'}^{N-1} a_{u,k}\Omega_{k,k'}^{u,u'}[\Delta_t]}_{\text{IS-ICI}} + \underbrace{w_{u',k'}}_{\text{Gaussian Noise}} \tag{7.52}$$

In Eq. (7.52) the first term stands for the attenuated useful signal, the second term denotes ISI, the third term gives IS-ICI and the last term stands for Gaussian noise.

7.4.2 Formulation of design problem

7.4.2.1 Design criterion

The information carried by the sub-carriers can be correctly decoded if the waveforms used have large *distances* between one another. In WPM this is achieved through the orthogonality of the generated waveforms. Therefore, in disturbance-free environments the cross-correlations of WPM waveforms equal zero and perfect reconstruction is possible despite the time and frequency overlap. The timing error Δ_t, on the other hand,

[12] The apostrophes in the symbol u' and carrier k' indices are used to indicate the receiver side.

leads to the loss of the orthogonality between the waveforms and consequently they begin to interfere one with another, leading to ICI and ISI, stated as:

$$\Omega^{u,u'}_{k,k';k\neq k'}[\Delta_t] = \sum_n \xi^k_l[n-uN]\xi^{k'}_{l'}[u'N-n+\Delta_t] \tag{7.53}$$

The design objective would therefore be to generate wavelet bases ξ and their duals ξ' that minimize interference energy in the presence of timing error:

$$\text{MINIMIZE:} \sum_{u,k;k\neq k'} \left|\Omega^{u,u'}_{k,k'}[\Delta_t]\right|^2 \quad \text{with respect to } \{\xi, \xi'\} \tag{7.54}$$

7.4.2.2 Wavelet domain to filter bank domain

The waveforms in WPM are created by the multi-layered tree structure filter bank. Using Parseval's theorem of energy conservation, it can be easily proven that the total energy at each level is equal regardless of the tree's depth. Therefore, minimizing the interfering energy at the roots of the tree will automatically lead to a decrease of total interfering energy at the higher tree branches. Furthermore, the two-channel filter banks through the 2-scale equation are related, albeit explicitly, to the WPM waveforms. Therefore, the design process can be converted into a tractable filter design problem. We should hence be able to minimize deleterious effects of time synchronization errors in WPM by minimizing the following cross-correlation function:

$$\sum_{\Delta_t} \left|r_{hg}[\Delta_t]\right|^2 = \sum_n |h[n]g[n-\Delta_t]|^2 = \sum_n \left|h[n]((-1)^n h[L-n+\Delta_t])\right|^2 \tag{7.55}$$

The design problem of minimizing the interference energy due to timing offset can now be formally stated as an optimization problem with objective function (7.55) and constraints (7.3), (7.5b) and (7.9), i.e.

$$\begin{aligned} &\text{MINIMIZE:} && \sum_{\Delta_t} \left|r_{hg}[\Delta_t]\right|^2 \quad \text{with respect to } h[n] \\ &&& \sum_n h[n] = \sqrt{2} \\ &\text{SUBJECT TO:} && \sum_n h[n]h[n-2k] = \delta[k] \quad \text{for } k = 0, 1, \ldots, (L/2)-1 \\ &&& \sum_n h[n](n)^k(-1)^n = 0 \quad \text{for } k = 0, 1, 2, \ldots, K-1 \end{aligned} \tag{7.56}$$

As in the first example, the majority of the constraints in Eq. (7.56) are non-linear and non-convex. And as before, we shall move to the autocorrelation domain ($r_h[k] = \sum_{m\in z} h[m]h[m+k]$) to simplify the problem.

7.4.3 Transformation of the mathematical constraints from a non-convex problem to a convex/linear one

The admissibility, paraunitary and K-regularity conditions are readily available in the autocorrelation domain (Eqs. (7.21), (7.23) and (7.28), respectively). The spectral factorization condition (7.36) can also be reused. Therefore, we only have to derive the objective function.

Now, we know that

$$r_h[n] = \begin{cases} \sum_{m=0}^{L-n-1} h[m]h[m+n] & n \geq 0 \\ r_h[-n] & n < 0 \end{cases} \tag{7.57}$$

and

$$\begin{aligned} r_g[n] &= \sum_{m=0}^{L-n-1} g[m]g[m+n] \quad \text{for } n \geq 0 \\ &= \sum_{m=0}^{L-n-1} ((-1)^m h[L-m])((-1)^{m+n} h[L-(m+n)]) = (-1)^n r_h[n] \end{aligned} \tag{7.58}$$

Applying the corollary[13]: *The sum of squares of a cross-correlation between two functions equals the inner product of the autocorrelation sequences of these two functions*, and the double shift orthogonality property:

$$r_h[2x] = \delta[x] = \begin{cases} 1, & \text{for } x = 0 \\ 0, & \text{otherwise} \end{cases} \quad \text{for } x = 0, 1, \ldots, \left\lfloor \frac{L-1}{2} \right\rfloor \tag{7.59}$$

the cross-correlation function $r_{hg}[n]$ can be rewritten in terms of $r_h[n]$ as follows:

$$\begin{aligned} \sum_{n=0}^{L-1} \left| r_{hg}[n] \right|^2 &= \sum_{n=0}^{L-1} r_h[n] r_g[n] \\ &= \sum_{n=0}^{L-1} r_h[n]((-1)^n r_h[n]) = \underbrace{\sum_{x=0}^{(L/2-1)} (r_h[2x+1])^2}_{\text{Odd numbered values}} - \underbrace{\sum_{x=0}^{(L/2-1)} (r_h[2x])^2}_{\text{Even numbered values}} \\ &= \sum_{n=0}^{(L/2-1)} (r_h[2n+1])^2 - 1 \end{aligned} \tag{7.60}$$

The new optimization problem can thus be stated as:

Minimize $\sum_{n=0}^{(L/2-1)} (r_h[2n+1])^2$ *subject to the wavelet constraints* (7.21), (7.23) and (7.28), *the spectral factorization criterion* (7.36) *and the design constraint*

[13] Proved in Appendix A3.

(7.60), i.e.

$$\begin{aligned}
&\text{MINIMIZE:} \quad \sum_{n=0}^{(L/2-1)} (r_h[2n+1])^2 \\
&\text{SUBJECT TO:} \quad \sum_{n=1}^{L-1} r_h[n] = \frac{1}{2} \\
&\qquad r_h[2k] = \delta[k] = \begin{cases} 1, & \text{for } k = 0 \\ 0, & \text{otherwise} \end{cases} \quad \text{where } k = 0, 1, \ldots, \left\lfloor \frac{L-1}{2} \right\rfloor \\
&\qquad \sum_{n=1}^{L-1} (-1)^n (l)^{2k} r_h[n] = 0 \quad \text{for } k = 0, 1, \ldots, K-1 \\
&\qquad r_h[0] + 2\sum_{n=1}^{L-1} r_h[n] \cos(i\pi l/d) \geq 0 \quad \text{for } i = 0, 1, \ldots, d.
\end{aligned} \tag{7.61}$$

The equations are convex and can be solved using the setup presented in Section 7.3.4 and illustrated by Figure 7.3.

7.4.4 Results and analysis

In this section we present a few results to demonstrate the design procedure [25]–[26]. As before, the main variables of the design process are the length and regularity order of the filter.

7.4.4.1 Frequency and impulse response of designed filter

In this example we have set the length of the filter to 20, though it is possible to design longer or shorter filters. The order of regularity chosen is 5, which is a compromise between optimization space and wavelet regularity. The impulse response of the designed optimal filter is illustrated in Figure 7.15 and numerical values of filter coefficients are given in Table 7.4. Although the optimal filter is designed in the autocorrelation domain, the minimum-phase time-domain coefficients obtained through spectral factorization satisfy all constraints mandated by the design process. The wavelet and scaling function of the newly designed optimal filter are illustrated in Figure 7.16a and b, respectively. The frequency response is shown in Figure 7.16c. Table 7.5 shows the specifications of the various filters used in the study along with the values of the corresponding objective functions. Clearly, the designed wavelet has the lowest interference energy.

7.4.4.2 Evaluation of designed filter under loss of time synchronization

The performance of the designed wavelet is compared and contrasted with several known wavelets by means of computer simulations. We have designed a communication system with DQPSK modulation and 128 orthogonal sub-carriers, corresponding to a wavelet packet tree of 7 stages. Guard intervals are not used and no error estimation or correction capabilities are implemented. To simplify the analysis, perfect frequency and phase synchronization are assumed. The time offset Δ_t is modelled as discrete uniform distribution between –2 and 2 samples, i.e., $\Delta_t \in [-2, -1, 0, 1, 2]$. In order to accentuate

Table 7.4 Optimal filter coefficients

0.119881851613898	0.498287367999060	0.660946808777660	0.203191803677134
–0.02915169068882	0.159448121842196	–0.144908809151642	–0.301681615791117
0.206305798368833	0.205999004857997	0.165410385138750	–0.0566148032177797
0.071282862607634	–0.00958254794419582	–0.0083940508446907	0.00912479119304040
–0.00498653232060	–0.00069440881953843	0.00154092796305564	–0.000370932610232259

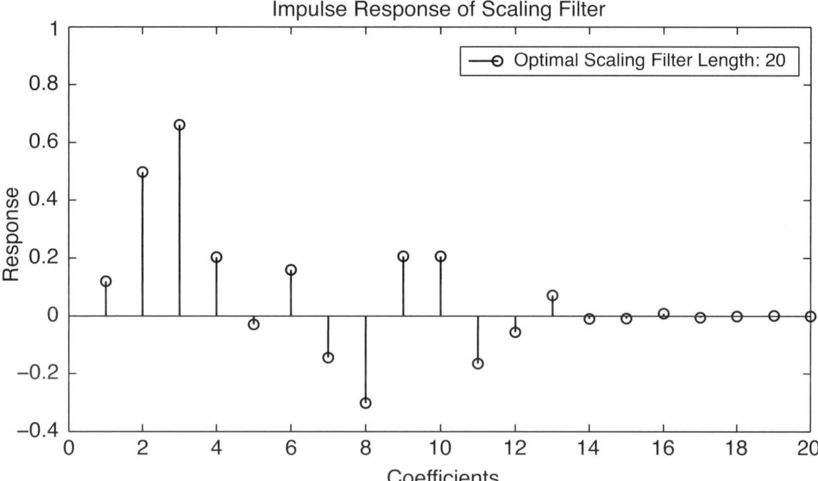

Figure 7.15 Impulse response of the optimal LPF with 20 coefficients.

the difference in performances between various wavelets, an oversampling factor of 15 is applied. The details are tabulated in Table 7.6.

Figure 7.17 shows the bit error rate (BER) performances of WPM system with different kind of wavelets and OFDM, under the assumption of an AWGN channel and time discrepancy between transmitter and receiver.

The plots in Figure 7.17 reveal that the designed optimal wavelet has better BER performance in the presence of timing error when compared to the performances of *standard* wavelets. However, OFDM tolerates a lot of time synchrony better than WPM. This is due to the fact that under time synchronization errors the ISI in OFDM arises only between two adjacent symbols, while in WPM several symbols interfere with each other. Table 7.7 shows the relative gains in the SNR performance of the designed optimal wavelet over *standard* wavelets in the presence of timing errors. The values have been calculated for a bit error rate (BER) of 10^{-4}.

BER and mean square error (MSE) calculated for different values of time offset are shown in Figures 7.18 and 7.19, respectively. Because the direction of timing error is inconsequential for a WPM-based system the time offset Δ_t is considered to follow a uniform distribution between 1 and 5 samples. The results presented corroborate the gains brought in by the newly designed wavelets.

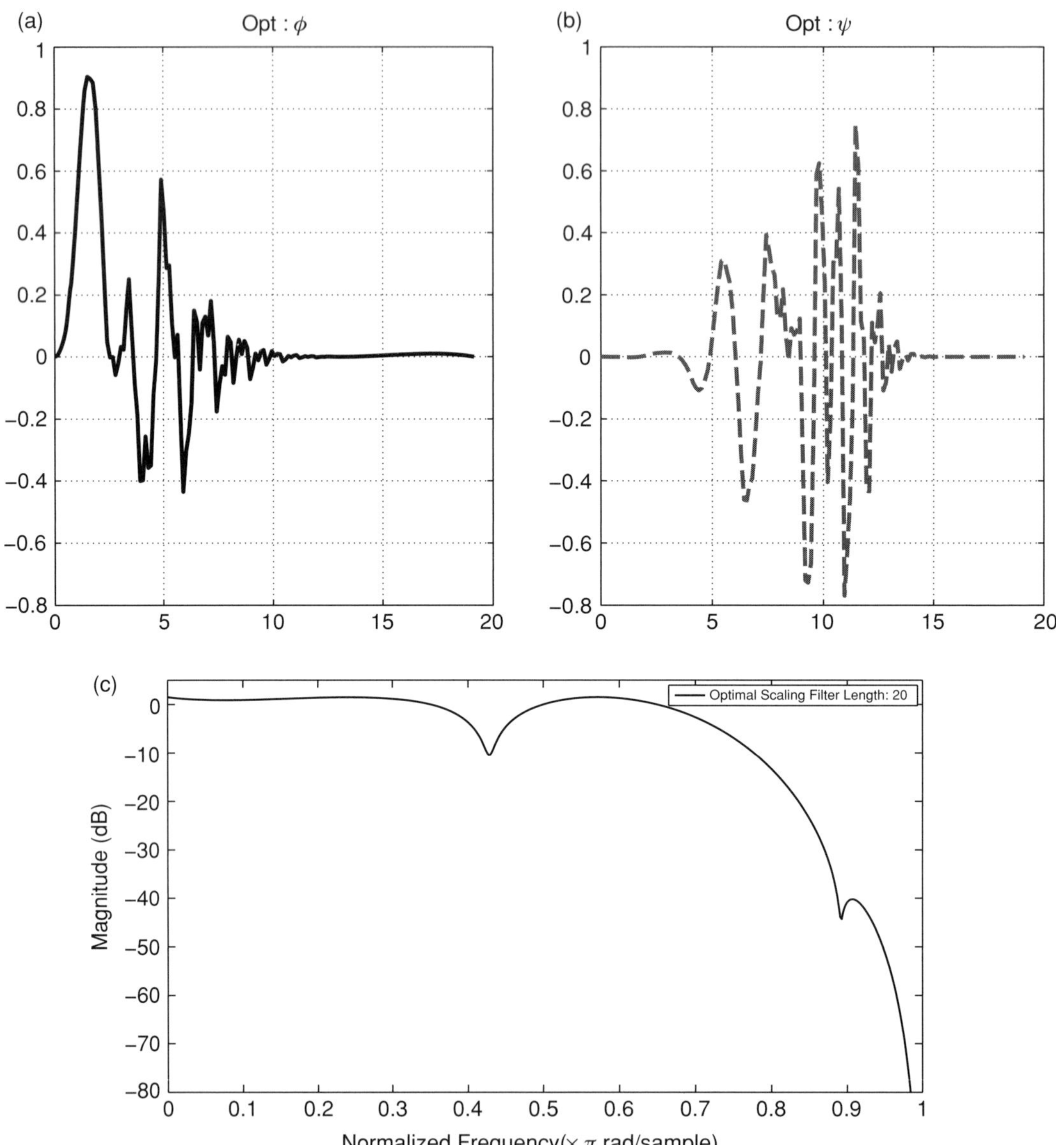

Figure 7.16 Optimal filter; (a) scaling function, (b) wavelet function and (c) frequency response in dB of the optimal filter.

7.4.4.3 Dispersion of sub-carrier energy

In Figures 7.20 (2D plots) and 7.21 (3D plots) the diffusion of sub-carrier energy to adjacent regions due to time synchronization error is portrayed. For clarity of depiction we have limited the number of sub-carriers to 16 and the WPM frame size to 30 multi-carrier symbols. From a total 480 sub-carriers in each frame, one pilot sub-carrier is set to a non-zero value, while the remaining 479 sub-carriers are made zero. The channel is

Table 7.5 Wavelet specifications and objective function

Name	Length	K-Regularity	$\sum_{n=0}^{(L/2-1)} (r_h[2n+1])^2$
Haar	2	1	–
Daubechies	20	10	0.41955
Symlets	20	10	0.41955
Discrete Meyer	102	1	0.45722
Coiflet	24	4	0.41343
Optimal	20	5	0.36814

Table 7.6 Simulation setup time synchronization error

	WPMCM	OFDM
Number of subcarriers	128	128
Number of multi-carrier symbols per frame	100	100
Modulation	DQPSK	DQPSK
Channel	AWGN	AWGN
Oversampling factor	15	15
Guard band	–	–
Guard interval	–	–
Frequency offset	–	–
Phase noise	–	–
Time offset	$\Delta_t = 2$	$\Delta_t = 2$

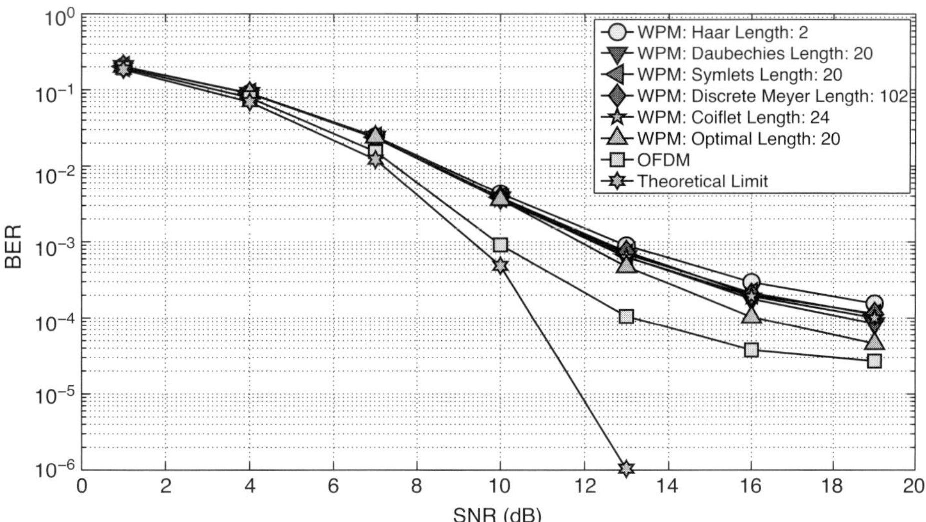

Figure 7.17 BER performance of different wavelets and OFDM under time synchronization errors.

Table 7.7 Performance improvement of designed optimal wavelet over standard wavelets in presence of timing errors (measured at BER of 10^{-4})

Wavelet Name	Haar	Daubechies	Symlets	Discrete Meyer	Coiflet
Improvement	5.03 dB	2.17 dB	3.25 dB	3.25 dB	2.98 dB

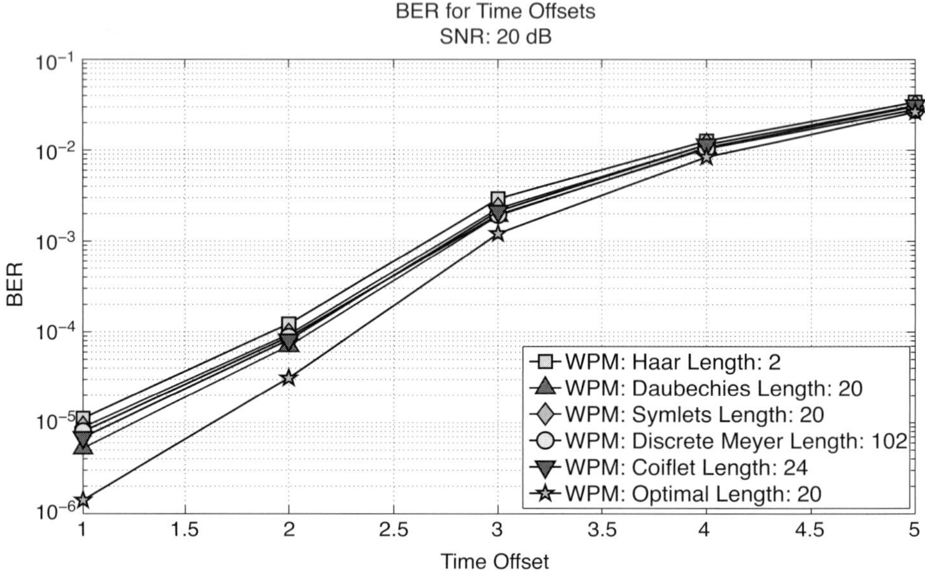

Figure 7.18 BER vs. time offset for WPM in AWGN channel (SNR = 20 dB).

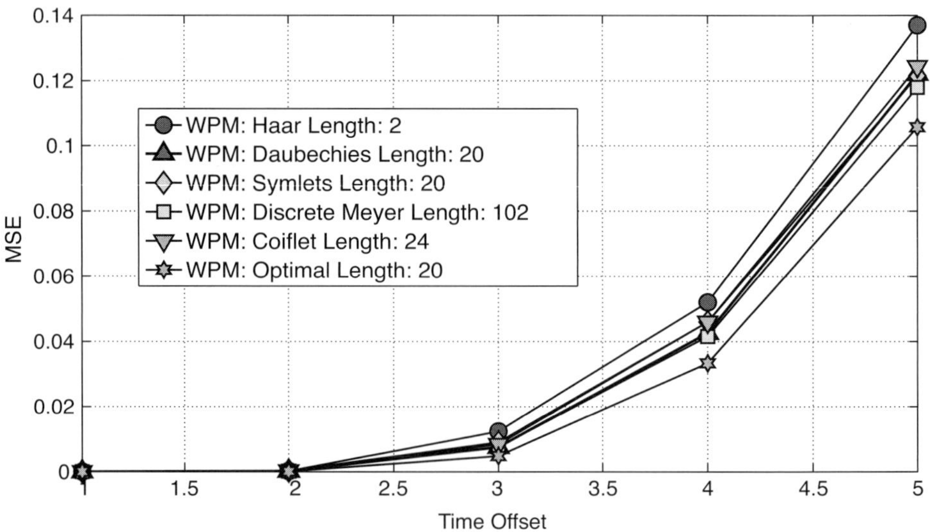

Figure 7.19 MSE vs. time offset for WPM in AWGN channel (SNR = 20 dB).

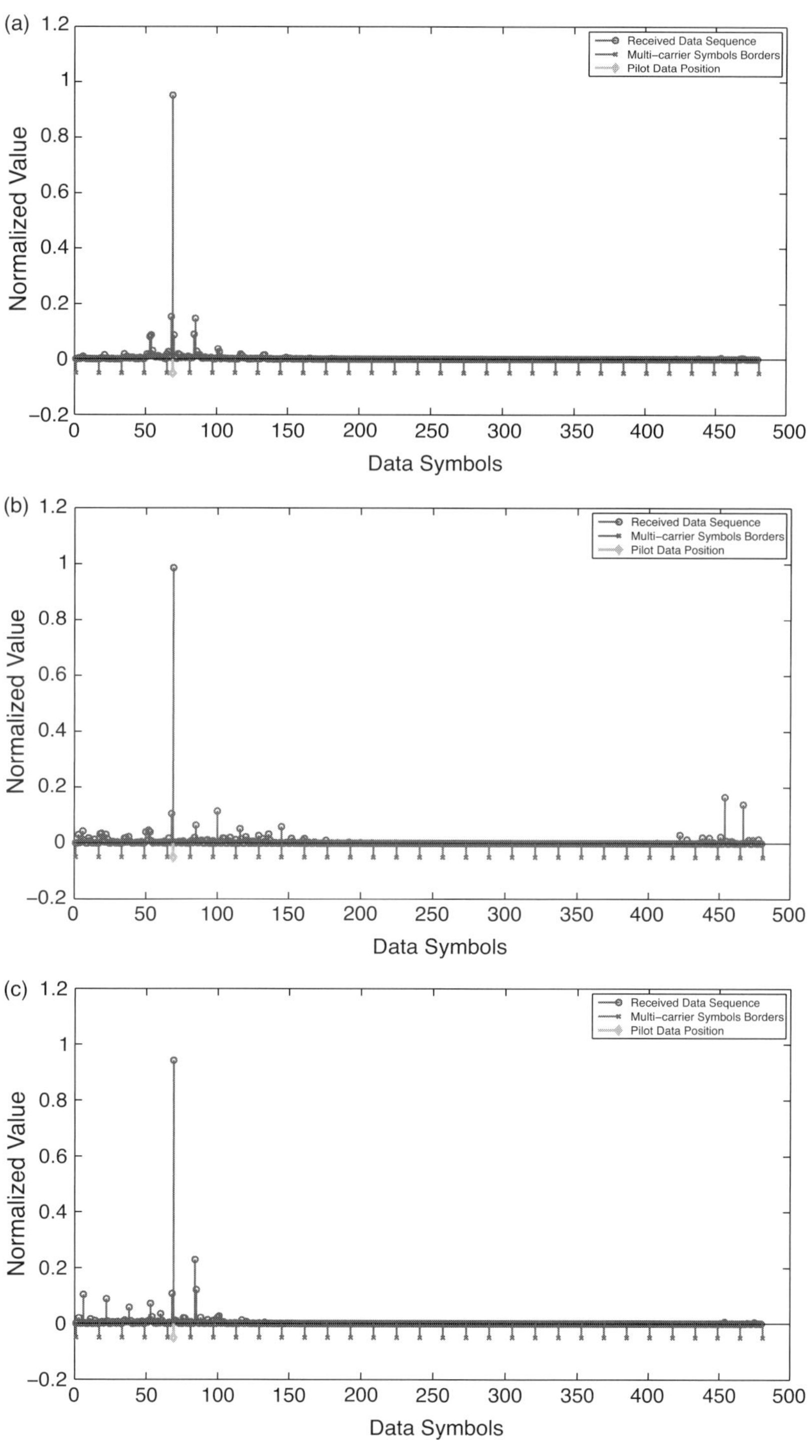

Figure 7.20 Received spectral energy (2D) in a frame in presence of timing error; a) Haar, b) Daubechies 20, c) Symlet 20, d) Discrete Meyer, e) Coiflet and f) optimal wavelet.

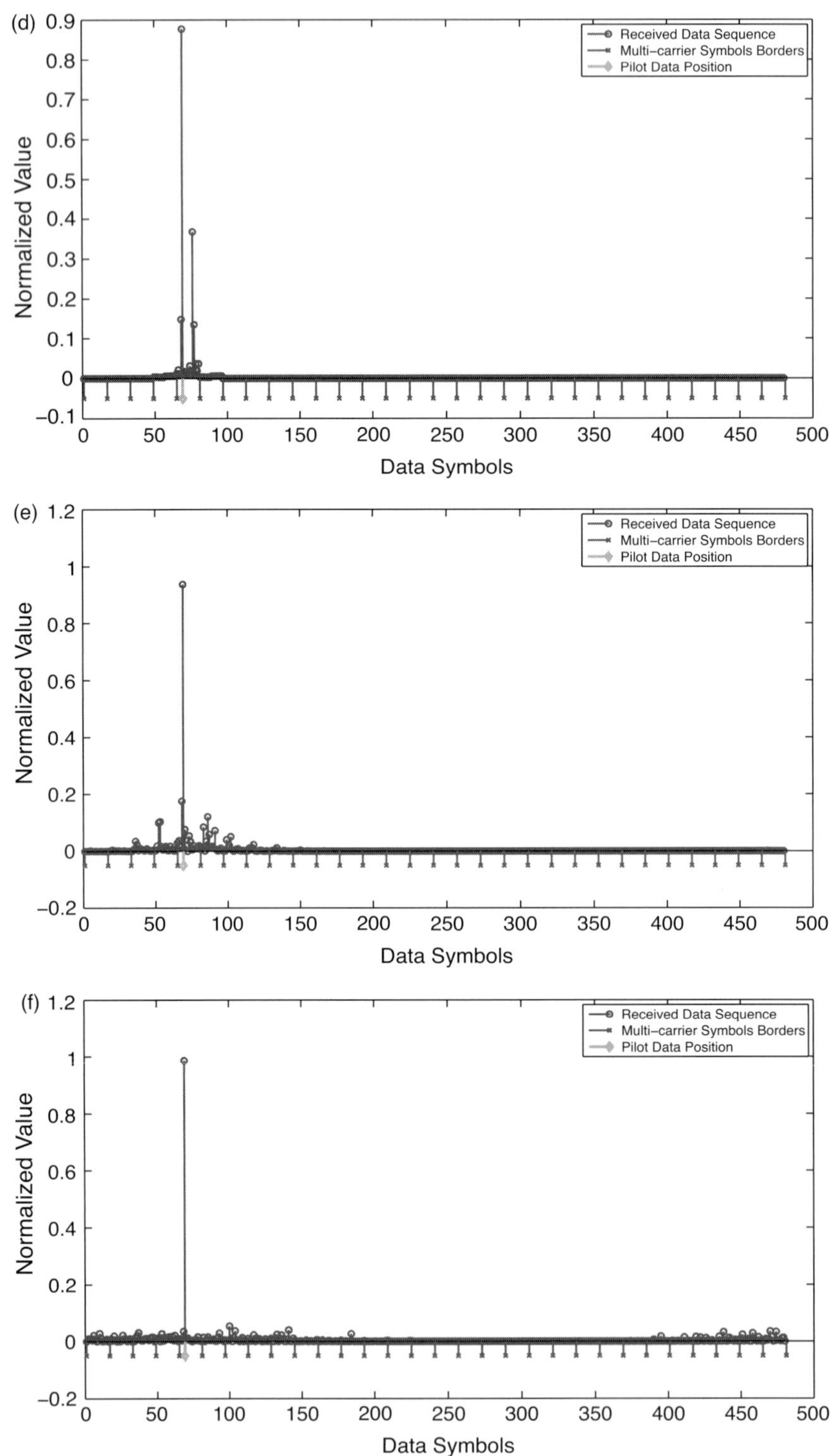

Figure 7.20 *(cont.)*

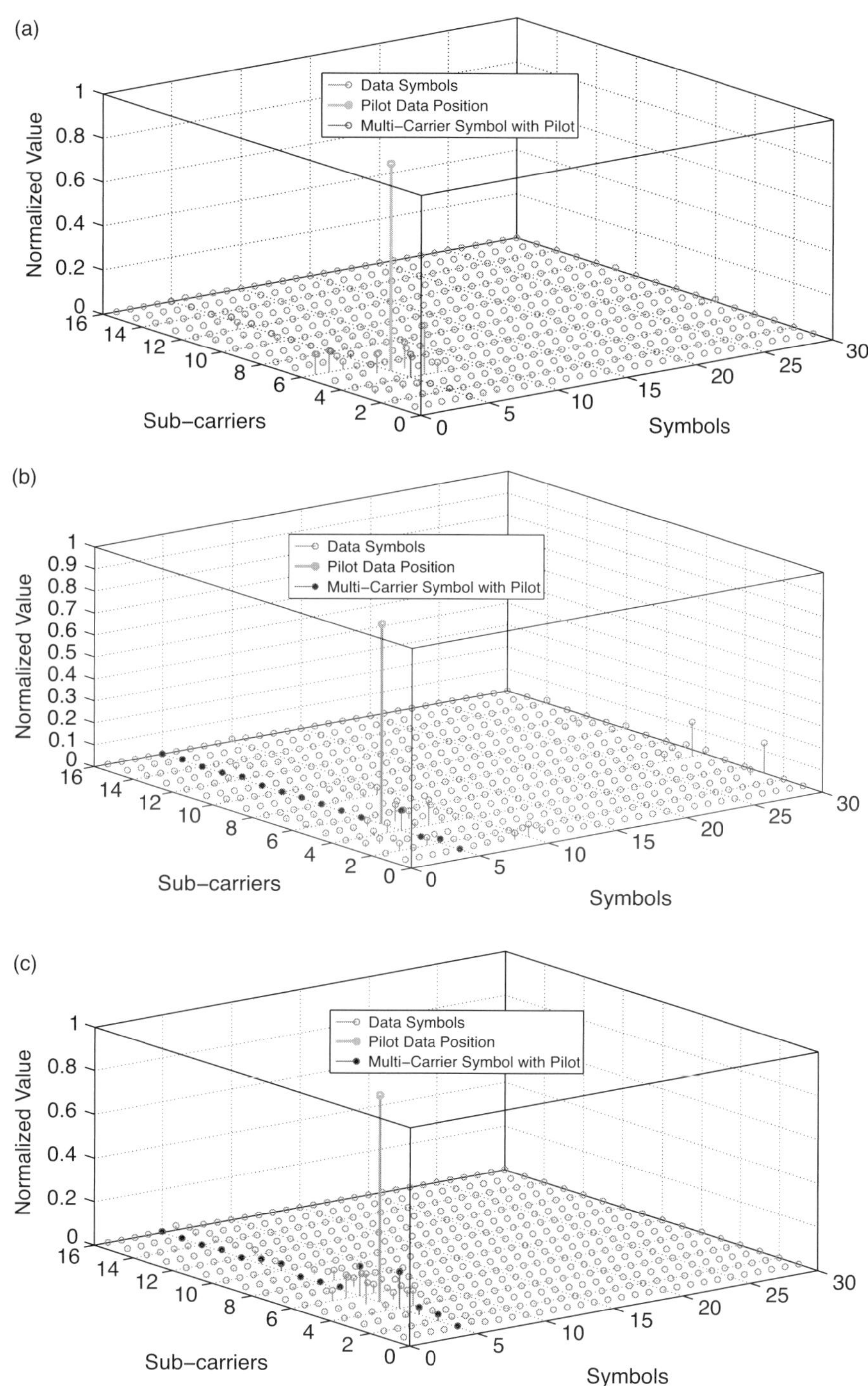

Figure 7.21 Received spectral energy (3D) in a frame in presence of timing error; a) Haar, b) Daubechies 20, c) Symlet 20, d) Discrete Meyer, e) Coiflet and f) optimal wavelet.

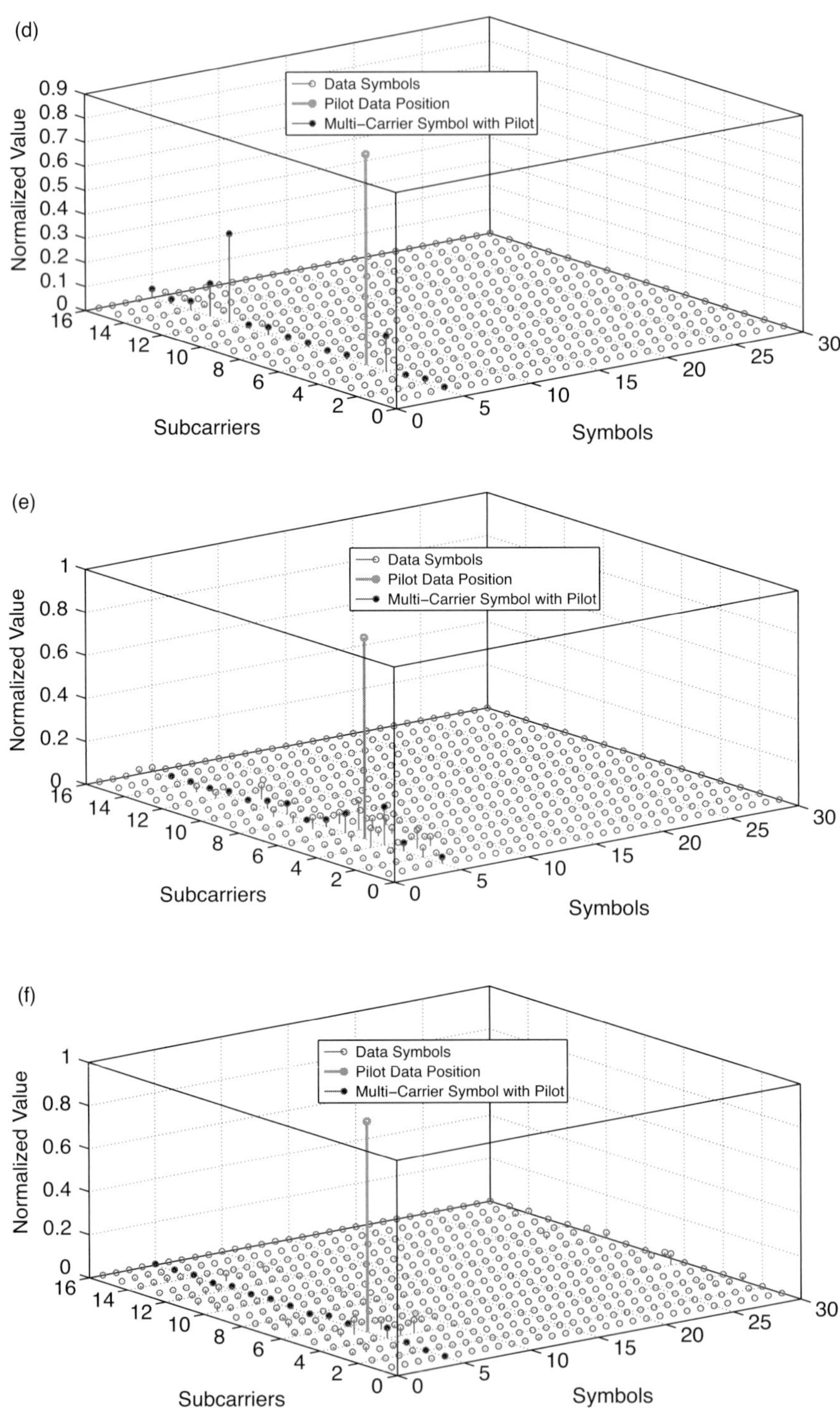

Figure 7.21 *(cont.)*

Table 7.8 Interference variance and maximum (max.) interference amplitude ratio

Wavelet Name	Haar	Daubechies	Symlets	Discrete Meyer	Coiflet	Optimal Wavelet
Interference Variance (10^{-5})	5.1088	3.0751	3.0721	3.1651	3.0639	2.8775
Max. Interference to Signal Amplitude	14.14%	8.36%	7.97%	6.66%	8.25%	4.51%

assumed to be ideal in order to emphasize the effect of timing error. It can be seen from the figures that the amplitudes of interfering sub-carriers is reduced by employing the newly designed filter (denoted 'optimal filter' in the graphs).

A note on the performance of the Haar wavelet: For the Haar wavelet the ISI is only limited to the neighbouring symbols, comparable to the OFDM case. This is because, in contrast to other orthogonal wavelets, waveforms generated by the Haar wavelet do not overlap in the time domain. However, despite the symbols not overlapping there are no gains to be had; this is due to increased ICI.

In Table 7.8 the values of interference variance and maximal interference amplitude ratio in relation to signal amplitude is given. These values are obtained for constant time offset and one pilot sub-carrier.

7.5 Summary

Wavelet packet modulation has recently emerged as a strong candidate for multi-carrier systems because of its offer of enormous adaptability and flexibility to system designers. In this chapter, we presented a methodology to design new wavelets according to a given design specification. The design process was described as an optimization problem that accommodated the design objectives and additional constraints necessary to ensure wavelet existence and orthonormality. In order to obtain the global minimum, the original non-convex constraints and objective function were translated into the autocorrelation domain. Using the new formulation, the design problem was expressed as a convex optimization problem and efficiently solved using a semi-definite programming technique. Two case studies – (a) where the cross-correlation between the filters was lowest, and (b) where the filters were maximally frequency selective – were used to demonstrate the design mechanism. The simulation results revealed that the newly designed wavelet satisfied all the design objectives and outperformed standard wavelets.

The wavelet design framework presented in this chapter can easily be applied for other design criteria (say reduction of PAPR or ISI or ICI) by merely altering the objective function. However, to be able to do so, the desirable properties of the wavelet bases must be translated into realizable objective functions. This can at times be challenging because the relationship between wavelet functions and filters is implicit and not direct. Another area of future research is to establish weights to gauge the trade-offs between various desirable (and at times contradictory) goals.

References

[1] A. Lindsay, "Wavelet Packet Modulation for Orthogonally Transmultiplexed Communications," *IEEE Transactions on Signal Processing*, vol. 45, pp. 1336–9, May 1997.

[2] I. Daubechies, "*Ten Lectures on Wavelets*," Society for Industrial and Applied Mathematics, Philadelphia, Pennsylvania, 1992.

[3] P.P. Vaidyanathan, *Multirate Systems and Filter Banks*, Prentice Hall PTR, Englewood Cliffs, New Jersey, 1993.

[4] W.J. Phillips, "Wavelet and Filter Banks Course Notes," 2003 [Available Online] http://www.engmath.dal.ca/courses/engm6610/notes/notes.html

[5] M. Vetterli and I. Kovacevic, *Wavelets and Subband Coding*, Englewood Cliffs, New Jersey: Prentice-Hall PTR, 1995.

[6] T.W. Parks and J.H. McClellan, "Chebyshev Approximation for Nonrecursive Digital Filters with Linear Phase," *IEEE Transaction. on Circuit Theory*, vol. CT-19, no. 2, pp. 189–94, March 1972.

[7] O. Rioul and P. Duhamel, "A Remez Exchange Algorithm for Orthonormal Wavelets," *IEEE Transactions on Circuits Systems – II*, vol. 41, no. 8, pp. 550–60, August 1994.

[8] E.L. Lawler and D.E. Wood, "Branch-And-Bound Methods: A Survey," *JSTOR Operations Research*, vol. 14, No. 4, pp. 699–719, August 1966.

[9] E. Alizadeh, "Interior Point Methods in Semidefinite Programming with Applications to Combinatorial Optimization," *SIAM Journal of Optimization*, vol. 5, pp. 13–51, 1995.

[10] Y. Ye, *Interior Point Algorithms: Theory and Analysis*, New York, Wiley, 1997.

[11] S. Boyd, L. El Ghaoui, E. Feron, and V. Balkrishnan, "Linear Matrix Inequalities in System and Control Theory," *SIAM Study in Applied Mathematics*, vol. 15, June 1994.

[12] S. Boyd and L. Vandenberghe, *Convex Optimization*, Cambridge University Press, 2004.

[13] L. Vandenberghe and S. Boyd, "Semidefinite Programming," *SIAM Review*, vol. 38, No. 1, pp. 49–95, March 1998.

[14] H. Wolkowicz, R. Saigal, and L. Vandenberghe, *Handbook of Semidefinite Programming*, Kluwer Academic Publisher, 2000.

[15] R. Hettich and K.O. Kortanek, "Semidefinite Programming: Theory, Methods and Applications, *SIAM Review*, vol. 35, No. 3, pp. 380–429, September 1993.

[16] A. Karmakar, A. Kumar, and R.K. Patney, "Design of an Optimal Two-Channel Orthogonal Filterbank Using Semidefinite Programming," *IEEE Signal Processing Letters*, vol. 14, No. 10, pp. 692–4, October 2007.

[17] J. Wu and K.M. Wong, "Wavelet Packet Division Multiplexing and Wavelet Packed Design Under Timing Error Effects," *IEEE Transaction on Signal Processing*, vol. 45, No. 12, pp. 2877–90, December 1997.

[18] J.K. Zhang, T.N. Davidson, and K.M. Wong, "Efficient Design of Orthonormal Wavelet Bases for Signal Representation," *IEEE Transactions on Signal Processing*, vol. 52, No. 7, July 2007.

[19] S.P. Wu, S. Boyd, and L. Vandenberghe, "FIR Filter Design via Semidefinite Programming and Spectral Factorization," in *Proceedings IEEE Conference on Decision and Control*, pp. 271–9, 1996.

[20] B. Alkire and L. Vandenberghe, "Convex Optimization Problems Involving Finite Autocorrelation Sequences," *Mathematical Programming*, Series A93, pp. 331–59, 2002.

[21] T.N. Davidson, L. Zhi-Quan, and K.M. Wong, "Orthogonal Pulse Shape Design via Semidefinite Programming," *IEEE International Conference on Acoustics, Speech and Signal Processing (ICASSP '99)*, vol. 5, pp. 2651–4, March 1999.

[22] T.N. Davidson, L. Zhi-Quan, and J.F. Sturm, "Linear Matrix Inequality Formulation of Spectral Mask Constraints With Applications to FIR Filter Design," *IEEE Transaction on Communications*, vol. 50, No. 11, pp. 2702–15, November 1999.

[23] J.F. Sturm, "Using SeDuMi 1.02, a MATLAB Toolbox for Optimization over Symmetric Cones," *Optimization Methods and Software*, vol. 11–12, pp. 625–53, 1999. [Available Online] http://sedumi.mcmaster.ca/

[24] J. Löfberg, "A Toolbox for Modeling and Optimization in MATLAB," *Proceedings of the CACSD Conference*, 2004. [Available Online] http://control.ee.ethz.ch/~joloef/wiki/pmwiki.php

[25] D. Karamehmedovic, M.K. Lakshmanan, and H. Nikookar, "Optimal Wavelet Design for Multicarrier Modulation with Time Synchronization Error," in *52nd Global Communications Conference (GLOBECOM)*, Hawaii, November 2009.

[26] M. Lakshmanan and H. Nikookar, "Construction of Optimal Bases for Wavelet Packet Modulation under Phase Offset Errors," in *3rd European Wireless Technology Conference*, Paris, France, September 2010.

[27] D.D. Ariananda, M.K. Lakshmanan, and H. Nikookar, "Design of Best Wavelet Packet Bases for Spectrum Estimation," *20th Symposium of Personal, Indoor and Mobile Radio Communications (PIMRC)*, Tokyo, Japan, September 2009.

[28] M. Lakshmanan, H. Nikookar, D. Karamehmedovic, and D.D. Ariananda, "A Unified Framework to Design Orthonormal Wavelets Packet Bases for Multi-Carrier Modulation," in *Proc. IEEE International Conference on Communications (ICC)*, Cape Town, South Africa, May 2010.

8 Conclusion

The convergence of information, multi-media and wireless communications has raised the vision of ubiquitous and pervasive communication – communication *anywhere, anytime and anything*. However, each of the incumbent wireless systems operates with different technologies, standards, interfaces, hardware (processors, radio frequency (RF) front end, antennas), software (drivers, firmware), network subscriptions, frequency bands and identities. In fact, state-of-the-art multi-radios only share the display and keyboard!

The consequence is that users have to constantly switch between different devices, modes and networks, impoverishing their experience. Therefore, there is an emergent need for a generic, universal radio that integrates different standards and air interfaces. The challenge is to seize on the right technical strategy to provide a common telecommunications medium that connects devices and thereby people.

Digital communication systems can be viewed as trans-multiplexers characterized by the transmission waveforms. The time–frequency properties of the pulse-shaping filter, i.e., the time spread and frequency footprint, determine the type of communication system (TDMA, FDMA, CDMA, OFDM, UWB, MC-CDMA, etc.). Different radios have different transmission characteristics that are greatly altered by the nature of the waveforms used. Transmission waveforms can thus be considered as the genes of the radios – the fundamental unit of change. Different manifestations of the waveforms can result in different systems with multifarious properties that can be tailored to different use cases. In order to integrate various systems we propose the realization of a flexible and generic wavelet-packet-based multi-carrier modulation (WPM) radio that can be adaptively reconfigured to maximize resource utilization and interference mitigation. Wavelets and wavelet transforms are used as the technology of choice because their characteristics can be widely customized to fulfil the requirements of intelligent wireless communication systems.

A generic wavelet-based MCM radio can emulate any signal. It can indeed be a natural replacement for multi-tone schemes like the popular orthogonal frequency division multiplexing (OFDM). It can also generate single-carrier schemes since they are just a special case of multi-carrier signals. A variant of UWB systems called multi-band OFDM (MB-OFDM) already exists where the wideband OFDM operates at different frequency bands at different instances. Multi-band-WPM can be an extension to MB-OFDM and can be applied for UWB transmission. Multiple access communication is also possible with wavelets. With their offer of greater flexibility in designing signature waveforms, and their inherent orthogonality property, they can play a vital role in the design of waveforms and receivers for multiple access systems.

The possibility of applying wavelet theory for the design of flexible and generic radios capable of handling multiple radios has been explored in the framework of this book. The results of these studies have been recorded in various publications and reports. Indeed, this book was dedicated to the demonstration of the wavelet packet modulator (WPM) as a wideband multi-carrier technology alternative to the well-known OFDM. Additionally, the application of WPM to a multi-antenna/MIMO architecture is investigated in [1]. In [2], a method to shape Ultra-wideband (UWB) signals using wavelet transform is presented. Bit error rate (BER) and outage probability performance of the proposed system in the presence of competing sources is analyzed and suitable strategies to mitigate the impact of interferers presented. And in [3], a novel receiver design that utilizes the time and frequency localization properties of wavelet transform is proposed for a wavelet-based single-carrier system.

It is important to underline here that all the modulation modes explained above employ the same signal processing architecture and hence they can be combined into a single radio unit.

The advantages of wavelet transform in terms of the flexibility they offer to customize and shape the characteristics of the waveforms have been demonstrated in [4]–[6]. Two use cases where the waveforms are designed and applied to optimize the system performance according to specific system demands are illustrated in [4]–[5]. In [6], the work of [4]–[5] is extended to establish a unifying mathematical framework where the waveforms can be designed according to any engineering requirement. Finally, the ability of wavelet radios to opportunistically exploit radio resources is illustrated in [7] where a WPM-based scheme for cognitive radio systems is addressed. In this proposal, the transmission waveform of WPM is sculpted to make use of the unoccupied time–frequency gaps of the licensed users.

In this book the operation of the novel wavelet packet modulator was evaluated. WPM is a flexible and adaptable radio that can be customized to a given user demand or engineering requirement. The challenges involved in the practical implementation of the system were listed and studied. The functioning of the proposed WPM system under various performance metrics was studied. Some of the figures of merit include:

- PAPR performance;
- sensitivity to loss of synchronization (time/frequency/phase);
- robustness to channel vagaries; and
- operation under interfering sources.

Suitable interventions that addressed issues like PAPR were devised, implemented and tested. The proposed system was successfully applied and verified for practical applications such as spectrum estimation and dynamic spectrum access. The adaptable features of the system were demonstrated in the form of wavelet design suited to the system specification.

Numerical results and comparative studies with FFT/OFDM based systems validated the efficacy of the algorithms deployed. The results of the study made it clear that wavelet radio can be a viable alternative for existing technologies.

The key inferences and conclusions of the study are summarized in the following sections.

8.1 Study of wavelet radio performance under loss of synchronization

- Orthogonal multiplexing schemes like WPM and OFDM are vulnerable to loss of synchronization in time, frequency or phase.
- Effect of frequency offset: The effect of frequency offset is to cause the sub-carriers to lose their mutual orthogonality, which results in mutual interference. In OFDM, the performance degradation is limited to the interference amongst the sub-carriers (referred as inter-carrier interference (ICI)) within one OFDM symbol duration. However, in WPM sub-carriers from multiple symbols interfere with each other causing inter-symbol–inter-carrier interference or IS–ICI. This dissimilarity in the interference behaviour is due to the manner in which the sub-carriers in wavelet- and Fourier-based systems are created. The signals generated by OFDM overlap only in the frequency domain while WPM generated signals overlap in both frequency and time domains.
- Impact of phase noise: Depending on the bandwidth of the phase noise, two scenarios can emerge in the presence of phase noise:
  - If the phase-noise bandwidth is small compared to the inter-carrier spacing, the dominant effect is a constant rotation of constellation symbols.
  - On the other hand, when the phase-noise bandwidth is greater than the inter-carrier spacing the rotational behaviour is less pronounced but instead the interference dominates.
- As with the effect of frequency offset, the interference due to the phase noise corrupts the OFDM signal only with ICI, while in WPM signals are corrupted with IS–ICI.
- Impact of loss of time synchronization: OFDM benefits from the use of cyclic prefix to greatly reduce the errors due to loss of time synchronization. WPM cannot benefit from such constructions due to time overlap of the symbols. Nevertheless, cyclic prefix in OFDM fails to prevent interference from occurring if the offset value is larger than the size of the prefix or when offset is in the direction opposite to the symbol's own prefix. When parts of the neighbouring symbols get erroneously selected at the OFDM or WPM receiver windows, the demodulated data are afflicted by ISI and ICI. In OFDM, ISI arises only due to neighbouring multi-carrier symbols, while in WPM more symbols, in addition to the contiguous ones, contribute to the generation of ISI.
- To understand the impact of loss of synchronization, analytical expressions were derived. To corroborate the theoretical findings, a computer simulation platform was set up and the performances of OFDM and WPM systems were examined in the presence of carrier frequency offset, phase noise and time synchronization error. Several well-known wavelets such as Daubechies, Symlets, discrete Meyer, Coiflets and biorthogonal wavelets were applied and studied.

- Simulations studies showed that the performance degradation of WPM and OFDM under carrier frequency offset and phase noise were comparable. However, the effect of time offset errors in WPM transceivers was far more severe than in OFDM.

8.2 PAPR performance studies

- The statistical distribution of WPM signal and its power variations were studied. The envelope of the WPM signal and its power were found to follow the Gaussian and chi-squared distribution, respectively.
- The effect of PAPR on the wavelet packet modulator (WPM) scheme was then evaluated. Various WPM configurations, with different wavelet families, pulse shapes and lengths, were considered for the study. OFDM was also included as reference. Almost all the wavelets performed similarly with regard to their PAPR performances. Moreover, the WPM operations were comparable with that of OFDM.
- To alleviate the PAPR impact on WPM transmission, three strategies can be employed: First, is a selected mapping (SLM) approach with phase modification technique to reduce the PAPR in the wavelet packet modulation system. By creating replicas of the original message by randomly altering the phases of the sub-carriers that modulate the information, different WPM frames with different PAPR values can be obtained. Then, the WPM frame with the least PAPR is transmitted. The attraction of this method is in its simplicity and elegance of implementation. Next is the selected mapping technique that can be implemented by optimizing the selection of phase offset of the sub-carriers. The SLM approaches necessitate the exchange of phase information between the transmitter and receiver ends. To overcome this, a third strategy, based on data scrambling, can be considered. In this method, replicas of the original input signal are suggested to be created by scrambling the original message with different scrambling sequences.

8.3 Wavelet-based spectrum sensing for cognitive radio

- A wavelet-packet-based spectrum estimator was developed. Since the mathematical precept of wavelets is tightly coupled to filter bank theory, the wavelet packet spectrum estimator (WPSE) was formulated as a filter bank analysis problem.
- An effectual spectrum estimator based on the theory of wavelets can be built by exploiting the filter bank's structure of wavelet packet decomposition.
- The decomposition level of the wavelet packet tree can be tuned to adjust the performance of the wavelet-based estimates with respect to variance of the estimated PSD and frequency resolution.
- Based on the level of decomposition, the WPSE performance ranges between that of Welch (i.e., shorter windowed periodogram) and periodogram.
- The wavelet-packet-based approach gives all wavelet coefficients at all decomposition levels. The presence of all of these coefficients allows for obtaining multiple estimates

from different levels of the tree with different degrees of variance and frequency resolution, in one snapshot and one operation. This feature can be exploited to construct an adaptable and reconfigurable spectrum estimation mechanism. This feature of WPSE can be of enormous advantage in a dynamic and variegating environment.

- The receiver operating characteristics (ROC) criteria was selected for the evaluation of the performance of wavelet-packet-based spectrum estimator. The variation of the performance of ROC was studied for different wavelet families (both orthogonal and non-orthogonal). The study of the variation of ROC with different wavelet filter properties like length of the wavelet filters, number of primary users and different signal-to-noise ratio of the input signal was also performed.
- Exploiting the prominent underutilization of the radio spectrum, an efficient wavelet-packet-based spectrum estimation model with reduced number of sensing measurements was proposed. The detection performance of the developed model was evaluated by extensive simulation. The developed model exhibits a significant reduction of number of sensing measurements with a promising detection performance.

8.4 Design of wavelets

We presented a general, unified approach to design and develop orthogonal wavelet packet bases according to a requirement. The important inferences of the study are detailed below:

- We advance the state-of-the-art in WPM to design wavelet bases for use in communication formats. This is necessitated by the fact that the wavelets currently in use are not custom-built for multi-carrier systems.
- To do so we estabished a generic, unified framework that facilitates the design of new wavelet bases that cater to a requirement.
- The possibility of adapting the characteristics of the WPM transmission is pursued with two examples where families of wavelets that are maximally frequency selective or have the lowest cross-correlation energy, are developed.
- The design of a wavelet filter is subject to the multiple constraints. Besides the primary goal (e.g., derivation of wavelets with high frequency selectivity) there are other constraints mandated by the wavelet theory that have to be fulfiled. These budgets of the filter design process were expressed as a convex optimization problem. The global solutions for the design problem were obtained using a mathematical numerical solver known as semi-definite programming (SDP). The solution filters were then tested for conformance with design goals.
- The WPSE results with the newly designed maximally frequency selective filters yielded more accurate results than ones based on standard wavelets such as Coiflets, Symlets and Daubechies.
- The WPM carrier using the new maximally frequency selective wavelets guaranteed sharper transition bands and better time–frequency localization than commonly known wavelets.

- To address the high sensitivity of WPM to time offset, we designed a new wavelet filter with low cross-correlation energy, which reduces the timing error interference.
- Studies on the WPM operation showed that the newly designed optimum filter ensured better performance under time synchronization errors when compared to standard wavelets such as Daubechies, Symlets, discrete Meyer, Coiflets, etc.

The results of these studies affirmed the promise that a WPM system holds in devising flexible communication systems whose characteristics can be tailored according to the engineering requirements.

8.5 Future research topics

8.5.1 Study of WPM performance under loss of synchronization

- In this book the vulnerability of WPM to time synchronization errors was addressed. However, in the implementation all the sub-carriers were taken to experience the same time or frequency offset. This model can be extended with different sub-carriers undergoing different offsets.
- Furthermore, a robust synchronization algorithm to detect and correct large time offsets can be implemented.

8.5.2 PAPR performance studies

- PAPR studies when the data are correlated (e.g., transmission of a picture or audio information) and explore its impact on the PAPR performance or on the mitigation techniques.
- Explore the possibility of data clipping as a PAPR mitigation technique and the utilization of wavelet de-noising methods at the receiver to retrieve data.
- Conduct PAPR studies with more sophisticated power amplifier models.
- Explore the possibility of designing new wavelets for PAPR reduction.
- Study the impact of oversampling on the PAPR performance of WPM.
- Exploit the tree structure of WPM to produce the best tree formation that guarantees minimum PAPR. Unlike OFDM that divides the communication channel into orthogonal sub-channels of equal bandwidths, WPM uses an arbitrary time–frequency plane tiling to create orthogonal subchannels of different bandwidths and symbol rates. When transmitting the same data in WPM, alternative tree representations can be used resulting in different sub-channel spacing in time and frequency that are not necessarily uniform. This feature of WPM can be utilized for the reduction of PAPR. In particular, each of the alternative (pruned) trees could result in a different value for PAPR, and an algorithm to choose the optimum tree structure can be devised such that the structure achieves the minimum PAPR.
- Complexity analysis study that considers the cost of implementing the reduction technique along with the loss in data rate.

8.5.3 Equalization of channel

- The studies in this book were confined to channels that were time invariant. Therefore, a natural extension to the work will be to consider channels that vary with time and/or are frequency dispersive. Frequency selective or dispersive channels can lead to a loss of orthogonality between signals of multi-carrier system causing disturbances such as inter-symbol interference (ISI) and inter-carrier interference (ICI). Channel equalization schemes should be used to encounter the deleterious effects of the wireless channel. Channel equalization for WPM is unique because the WPM symbols overlap in time. Hence, both ISI and inter-symbol–inter-carrier interference (IS–ICI) occur and have to be factored in the design of equalizer.
- A simple equalizer can be focused on the removal of ISI. Devising an equalizer that handles both the inter-symbol interference and inter-carrier interference can be a fruitful area of further research.
- A blind equalizer in which the transmitted signal is inferred from the received signal making use only of the transmitted signal statistics (without availability of channel information), can be productive area of research.
- Channel modelling and better representation of channel using wavelet packets. This information can be used to customize the transceiver tree structure based on the channel condition and further simplify the equalization process.

8.5.4 Wavelet packet spectrum estimator (WPSE)

- Explore the possibility of applying compressed sampling for wavelet-packet-based spectrum estimation.
- Expand edge mitigation studies to include more sophisticated approaches to reduce artifacts introduced windowing in WPSE filter bank implementation.
- Study of dual-tone sources to understand the ability of WPSE to resolution abilities of WPSE.
- Derive analytical expressions for WPSE variance and bias.
- Explore the possibility of applying windows to the WPSE method to tackle spurious spectral growth (also known as spectral carving).
- Co-operative spectrum sensing: Diversity exploitation to improve the WPSE probability of errors/estimation. This can also be useful in avoiding shadowing and hidden node problems.
- Analyze the ability of WPSE for sparse representation of the radio environment (frequency information) with lower number of coefficients. This helps in a co-operative spectrum sensing scenario where less information sharing indirectly leads to less clogging of the bandwidth.
- Investigation of the properties of WPSE to guarantee good time resolution. This property can be useful in scenarios where the information on time can be vital (e.g., estimation of swept tone sources).
- In [8] Tian and Giannakis propose a wavelet-based wideband spectrum sensing approach for dynamic spectrum management. In their approach, the signal

spectrum over a wide frequency band is decomposed into elementary building blocks of non-overlapping sub-bands that are well characterized by local irregularities in frequency. Then, the entire wideband is modelled as a sequence of consecutive frequency sub-bands, where the power spectral characteristic is smooth within each sub-band but exhibits a discontinuous change between adjacent sub-bands. Information on the locations and intensities of spectrum holes and occupied bands is derived by considering the irregularities in PSD. The main attraction for wavelets in this application is in their ability to analyze singularities and irregular structures that can be used to characterize the local regularity and edges of signals. Hence, the method is also called edge detection. The method of Tian and Giannakis can be enhanced further by designing new wavelets that are best suited for the application instead of using arbitrary wavelets.

- The receiver operating characteristic performance of WPSE/WPM system was evaluated for various wavelet filters. The study can be furthered to include more scenarios, in particular, the cases where the nature of the primary user varies frequently or when the data available to gauge the primary user characteristics is limited.
- Evaluation of system performance under different channel conditions.

8.5.5 Design of wavelets

- The unique features offered by wavelets to tailor and customize new filters were explored in this work to make WPM transmission less sensitive to time synchronization errors or reduce spectral spillage into neighbouring bands. The wavelet filter design template developed in this work can also be used for other design goals by merely altering the objective function and other design *budgets*. For instance, wavelet filters that can decrease frequency offset and phase noise sensitivity, reduce peak-to-average power ratio (PAPR) or increase spectral efficiency could be designed. However, to be able to do so, the desirable properties of the wavelet bases must be translated into realizable objective functions. This can at times be challenging because the relationship between wavelet functions and filters is implicit and not direct.
- Another area of research is to establish weights to gauge the trade-offs on offer between various desirable (and at times contradictory) goals.
- An added advantage of using wavelet theory for multi-carrier modulation is in the possibility of improving transmission security. Because newly designed wavelets are unique in nature, the transmitted signal can only be decoded by the WPM receiver that is acquainted with filter coefficients used by the WPM transmitter.

8.6 Related studies

One of the prime motives for pursuing wavelet-based systems is in the flexibility and adaptability that they offer. This capability can be readily exploited to provide better services to users and enrich their experience. Indeed, to realize these capabilities many technological challenges have to be overcome. Apart from that there are other related subjects that have to be addressed. Foremost amongst them is an understanding of how an engineering requirement translates to a particular system specification. To successfully map this relation, a careful and thorough study of the following areas must be conducted:

i. Impact of waveform characteristics on various performance metrics.
ii. The trade-offs on offer with regard to design of waveforms. (e.g., Does the waveform that yields best PAPR performance affect the BER performance? What about its performance with respect to ISI/ICI reduction?).
iii. A thorough analysis on the complexity issues.
iv. Mapping the complexity analysis and trade-off on offer in an easy to understand and presentable manner such that the user can make effective choices based on his needs and necessities. For example, if the user desires low battery power consumption he should be able to operate with a single radio with sufficient features for gainful communication. On the other hand, he should also be made aware of the consequences of his choice and how it may affect the quality of service (such as lower speed, lower bandwidth, throughput, etc.).

Other topics that have to be covered for the practical implementation of WPM include:

i. ensuring backward compatibility with existing technologies and systems;
ii. making the system generic and flexible so that it can be easily scaled;
iii. establishing suitable mechanisms to analyze radio environment and utilize them effectively in radio reconfiguration schemes;
iv. demonstration of the system capabilities through a proof of concept;
v. development of software tools to adequately test and verify the system;
vi. standardization of the technology to ensure compatibility across different development platforms.

8.7 Beyond this book

Apart from the suggested improvements catalogued above, there are other areas of wireless system design where the wavelet packet architecture can be readily applied. Here, we discuss a few of them.

8.7.1 Wavelet-based modelling of time-variant wireless channels

Currently available wireless channel models are based on statistical impulse response models derived from empirical results. While these models perform adequately for time-invariant channels, they fail to accurately map time-varying channels. Wavelet transform is a way of decomposing a signal of interest into a set of basis waveforms, called wavelets, which thus provide a way to analyze the signal by examining the coefficients (or weights) of wavelets. Due to their inherent joint time–frequency localization property and their ability to accurately characterize the time-varying nature of the estimation problem, the wavelets offer various advantages for channel modelling. Some of these are: accurate characterization of time-varying as well as frequency selective multi-path fading channels, fast convergence of estimate, representation of channel with fewer number of coefficients, small output error and clear interpretation of modelling error.

8.7.2 Multiple-access communication

Wavelets and wavelet packets possess unique properties that make them attractive for use in multiple-access communications. With their offer of greater flexibility in designing signature waveforms, and their inherent orthogonality property, they can play a vital role in the design of waveforms and receivers for multiple-access systems. Furthermore, wavelets can facilitate the design of user signature waveforms for code division multiple access (CDMA) communication systems. By randomly clipping the wavelet construction tree, a complete and orthonormal basis is generated. This basis eventually spawns spreading codes that are orthogonal to one another. Moreover, they display greater capacity to suppress multiple access interferences. The design and construction of orthogonal signatures for use in a spread signature code division multiple access system (CDMA) are discussed in [9]. According to [10], wavelets allow for simpler equalization and detection of CDMA signals at the receiver.

Multi-carrier CDMA or MC-CDMA is a data transmission technique that combines multi-carrier modulation (MCM) and CDMA. It is a spread spectrum technology, where the spreading is performed in the frequency domain, unlike CDMA, where the spreading is done in the time domain. By combining the best of MCM and CDMA, MC-CDMA promises high speed, large bandwidth, better frequency diversity to combat frequency-selective fading and good performance in severe multi-path conditions. MC-CDMA has thus emerged as a strong candidate for future wireless systems. In comparison to the conventional Fourier-based MC-CDMA systems, introducing wavelets to MC-CDMA yields the following advantages:

i. They provide three levels of orthogonality:
   - between the sub-carriers;
   - between the wavelets and scaling functions;
   - between the spreading sequences.

Therefore, in comparison to conventional MC-CDMA systems, they offer new dimensions to combat multi-path fading, ICI and interference or jamming signal by providing

ii. flexibility in choosing the spacing between the sub-carrier frequencies;
iii. flexibility in choosing the wavelet family based on need.

8.7.3 Wavelet radio for green communication

Recent studies have shown that the energy costs account for as much as half of a mobile service provider's annual operating expenses. Therefore, making the communication equipment more efficient in relation to its power consumption not only has implications with regard to environmental pollution but also makes economic sense. The theme of green communications is to design energy-efficient communication techniques and protocols that optimally utilize available resources and minimize power consumption. Wavelet-based technologies offer a lot of tools for research and development of green communication devices.

The methodologies on offer can be classified into two broad categories:

i. *Customization of waveforms*
   - While there are no explicit relationship between power optimization and waveforms, the nature and characteristics of the waveform can be altered to suit a set of requirements that can indirectly contribute to a more efficient system resulting in lower requirements of power and energy. These criterion could typically be:
     - minimization of ISI, ICI or PAPR;
     - greater tolerance and robustness to time/frequency/phase offset errors;
     - robustness towards interference from competing sources;
     - possibilities for opportunistic communication (e.g., cognitive radio) where unused resources can be cleverly utilized.
   - It is important to note that every performance metric that is influenced by the characteristics of the transmission waveform can be mapped into a design constraint and exploited to yield efficient systems.

ii. *Customization of tree structure*
   - The wavelet based systems are realized from a tree structure obtained by cascading a fundamental quadrature mirror filter (QMF) pair of low- and high-pass filters. The construction of this tree structure can be adjusted to produce an optimum tree structure that caters to various requirements. The requirements could typically be:
     - Identification and isolation of the "atoms" of interference in both time and frequency domains.
     - Flexibility with time–frequency tiling of the carriers that can lead to multi-rate systems that can transmit with different rates in different bands. This feature can be exploited in scenarios where the channel characteristics are not uniform.
   - It can be proven that the complexity of the wavelet systems is by and large simpler than OFDM systems. A lower complexity also means lower power requirements in the execution of the signal-processing algorithms. The implementation of wavelet systems can be simplified even further if fast wavelet transforms are employed.
   - In addition to these advantages, the promise of an integrated and universal wavelet-based radio can also immensely help in the optimization of the system performance. By integrating multiple radios, the wavelet-based systems remove the need for multiple firmwares, software, drivers, etc., and reduce power consumption and improve battery life.

8.7.4 Wavelet-based multiple-input–multiple-output communications (MIMO)

The OFDM–MIMO (multiple-input–multiple-output) combination has been successfully applied to enhance the throughput and range of wireless networks without the ensuant increase in bandwidth or power requirements. Since OFDM and WPM share many properties, as both are orthogonal multi-carrier techniques, there is potential for a WPM system to be used in a multi-antenna MIMO setup.

8.8 Concluding remarks

In a recent article in the Communication magazine, Steve Weinstein [11], a pioneer in the development of OFDM, traces back the journey of OFDM right from its inception in 1966 when Chang [12] published the first paper on multi-carrier modulation, to the development of the first proof of concept by Bell Labs in 1985 [13] and its first major consumer deployment as ADSL in 1993 and finally its standardization as IEEE 802.11a in 1999. And in his concluding remarks he advocates wavelet-based systems as true successors of OFDM, especially for the development of futuristic low-power *green radios* that are intelligent and adaptable.

The research and investigation on the utilization of wavelet technology for smart resource-aware radio systems, as presented in this book, can be considered as a fruitful attempt at tackling the various technical questions that will shorten the development time from conception to practical realization of wavelet radios. Furthermore, in an era when bold predictions that the *PHY Layer is Dead* [14] are made, the work on wavelet-based radios can increase the capacities of the wireless link and open new vistas for gainful research on radio design.

References

[1] M.K. Lakshmanan, I. Budiarjo and H. Nikookar, "Wavelet Packet Multi-carrier Modulation MIMO Based Cognitive Radio Systems with VBLAST Receiver Architecture," *IEEE Wireless Communications and Networking Conference (WCNC)*, March–April 2008, Las Vegas, USA.

[2] M.K. Lakshmanan and H. Nikookar, "Mitigation of Interference from Wideband IEEE 802.11a Source on UWB Wireless Communication using Frequency Selective Wavelet Packets," *International Waveform Diversity and Design (WDD) Conference*, June 2007, Pisa, Italy.

[3] M.K. Lakshmanan and H. Nikookar, "Best Bases Selection for Wavelet Domain Communication Systems," *International Waveform Diversity and Design (WDD) Conference*, 2009, Orlando, USA.

[4] D. Karamehmedovic, M.K. Lakshmanan, and H. Nikookar, "Optimal Wavelet Design for Multicarrier Modulation with Time Synchronization Error," *52nd IEEE Global Communications Conference (GLOBECOM)*, November 2009, Hawaii, USA.

[5] D.D. Ariananda, M.K. Lakshmanan, and H. Nikookar, "Design of Best Wavelet Packet Bases for Spectrum Estimation," *20th IEEE Symposium of Personal, Indoor and Mobile Radio Communications (PIMRC)*, Tokyo, Japan, September 2009.

[6] M.K. Lakshmanan, H. Nikookar, D. Karamehmedovic, and D.D. Ariananda, "A Unified Framework to Design Orthonormal Wavelets Packet Bases for Multi-Carrier Modulation,"

Proc. IEEE International Conference on Communications (ICC) 2010, Cape Town, South Africa, May 2010.

[7] M.K. Lakshmanan and H. Nikookar, "Construction of Optimum Wavelet Packets for Multi-carrier based Spectrum Pooling Systems," *Special Issue Wireless Personal Communications*, vol. 54, June 2010.

[8] Z. Tian and G.B. Giannakis, "A Wavelet Approach to Wideband Spectrum Sensing for Cognitive Radios," *Proceedings of International Conference on Cognitive Radio Oriented Wireless Networks and Communications*, Greece, June 8–10, 2006.

[9] G. Wornell," Emerging Applications of Multirate Signal Processing and Wavelets in Digital Communications," *Proc. IEEE*, vol. 84, no. 2,2, pp. 586–603, April 1996.

[10] H. Zhang and H. Fan, "Wavelet Packet Waveforms for Multicarrier CDMA Communication," *Proc. International Conf. on Acoust. Speech, Signal Proc.*, Orlando, FL, May 2002.

[11] S.B. Weinstein, "Introduction to the History of OFDM," *IEEE Communications Magazine*, November 2009, Volume 47, No. 11.

[12] R.W. Chang, "High-Speed Multichannel Data Transmission with Bandlimited Orthogonal Signals," *Bell Sys. Tech. J., vol.*45, Dec. 1966, pp. 1775–96; see also U.S. Patent 3,488,445, Jan. 6, 1970.

[13] "A Brief History of OFDM", 2010, Web link: http://www.wimax.com/commentary/wimax_weekly/sidebar-1-1-a-brief-history-of-ofdm

[14] Panel Discussions at IEEE 69th Vehicular Technology Conference (VTC-Spring), April 2009, Barcelona, Web link: http://www.ieeevtc.org/vtc2009spring/panels.php#Panel02

A1 Semi-definitive programming

Semi-definite programming (SDP) is a sub-field of convex optimization, which can efficiently exploit interior point methods to find an optimal solution. The main advantages of convex optimization methods are that they always achieve global minimum without being trapped in the local minima, and that they can determine explicitly the feasibility of a given set of constraints. SDP algorithms can be used to solve linear, quadratic and semi-definite problems, which all are part of convex optimization problems.

The optimization problems in SDP can be described as minimization of an objective linear function over the intersection of the semi-definite cone with an affine space. This cone is shaped by constraints that form a set of positive symmetric semi-definite matrices, called the linear matrix inequality (LMI) constraint [ref. 11 of Chapter 7]. LMI gives boundaries of feasible region in which SDP solver tries to find an optimal solution for the objective function or prove infeasibility. This region is generally non-smooth and non-linear but it has to be convex in order to be solvable by SDP [ref. 11 of Chapter 7]–[ref. 15 Chapter 7].

Some set C is said to be convex if the line segment between any two arbitrary selected points in C lies also in C. In case of points X_1 and X_2 we can show convexity by:

$$\Gamma X_1 + (1 - \Gamma)X_2 \in C \text{ for } X_1, X_2 \in C \text{ and } 0 \leq \Gamma \leq 1 \qquad (A1.1)$$

The example of convex and non-convex sets is illustrated by Figure A1.1. For each two points in the pentagon the line segment lies in the defined set and therefore the left figure is convex. The right figure is obviously not convex since two points X_1 and X_2 are within the set but the line that connects them is partially not contained in the set.

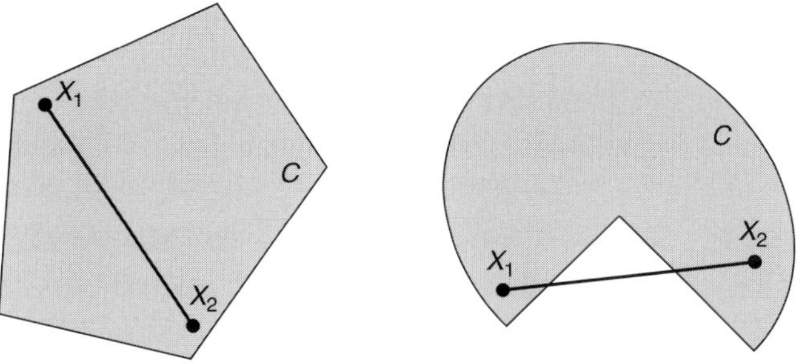

Figure A1.1 Convexity; left: convex set, right: non-convex set.

A2 Spectral factorization

The Kolmogorov spectral factorization method is based on construction of the minimum phase spectral factor $SF_{mp}(z)$ from the autocorrelation function. The power-series expansion of $SF_{mp}(z)$ is given by:

$$\log SF_{mp}(z) = \sum_{n=0}^{\infty} d_n z^{-n} \quad \text{for } z \geq 1 \tag{A2.1}$$

We can decompose $\log SF_{mp}$ into real and imaginary parts as:

$$\log SF_{mp}(z) = \mu(z) + j\eta(z) \tag{A2.2}$$

Here, $\mu(\omega)$ and $\eta(\omega)$ are Hilbert transform pairs. For $z = e^{j\omega}$ we have:

$$\begin{aligned} \mu(\omega) &= \log \left| SF_{mp}(e^{j\omega}) \right| \\ &= \frac{1}{2} \log R_h(\omega) \\ &= \sum_{n=0}^{\infty} d_n \cos n\omega, \; \eta(\omega) = -\sum_{n=0}^{\infty} d_n \sin \omega n \end{aligned} \tag{A2.3}$$

In Eq. (7.33) $R_h(\omega)$ denotes the Fourier transform of autocorrelation sequence. We can find the coefficients d_n by:

$$d_n = \frac{1}{2\pi} \int_0^{2\pi} \frac{1}{2} \log R_h(\omega) e^{-jn\omega} d\omega \tag{A2.4}$$

A3 Sum of squares of cross-correlation

The sum of squares of cross-correlation magnitude is related to the autocorrelation sequences of low-pass filter H and high-pass filter G according to the following equation:

$$\begin{aligned}\sum_{n=0}^{L-1} \left|r_{hg}[n]\right|^2 &= \sum_{n=0}^{L-1} r_h[n]\,((-1)^n r_h[n]) \\ &= \langle r_h[n], r_g[n]\rangle\end{aligned} \tag{A3.1}$$

Proof

$$\begin{aligned}\sum_n \left|r_{hg}(n)\right|^2 &= \sum_n \left(\sum_p h[p+n]g[p]\right)^2 \\ &= \sum_n \sum_m \sum_p h[p+n]g[p]h[m+n]g[m] \\ &= \sum_m \sum_p g[p]g[m] \sum_n^L h[p+n]h[m+n] \\ &= \sum_m \sum_p g[p]g[m] \sum_{n=m-p} h[m]h[2m-p] \\ &= \sum_m \sum_p r_h[m-p]g[p]g[m] \\ &= \sum_p \sum_{n=m-p} r_h[n]g[p]g[n+p] \\ &= \sum_n r_h[n]r_g[n] \\ &= \langle r_h[n], r_g[n]\rangle\end{aligned} \tag{A3.2}$$

Index